Thermodynamics of Heat Engines

SCIENCES

Energy
Field Directors – Alain Dollet and Pascal Brault

Physics of Energy and Energy Efficiency
Subject Head – Michel Feidt

Thermodynamics of Heat Engines

Coordinated by
Bernard Desmet

WILEY

First published 2022 in Great Britain and the United States by ISTE Ltd and John Wiley & Sons, Inc.

Apart from any fair dealing for the purposes of research or private study, or criticism or review, as permitted under the Copyright, Designs and Patents Act 1988, this publication may only be reproduced, stored or transmitted, in any form or by any means, with the prior permission in writing of the publishers, or in the case of reprographic reproduction in accordance with the terms and licenses issued by the CLA. Enquiries concerning reproduction outside these terms should be sent to the publishers at the undermentioned address:

ISTE Ltd
27-37 St George's Road
London SW19 4EU
UK

www.iste.co.uk

John Wiley & Sons, Inc.
111 River Street
Hoboken, NJ 07030
USA

www.wiley.com

© ISTE Ltd 2022

The rights of Bernard Desmet to be identified as the author of this work have been asserted by him in accordance with the Copyright, Designs and Patents Act 1988.

Any opinions, findings, and conclusions or recommendations expressed in this material are those of the author(s), contributor(s) or editor(s) and do not necessarily reflect the views of ISTE Group.

Library of Congress Control Number: 2022941652

British Library Cataloguing-in-Publication Data
A CIP record for this book is available from the British Library
ISBN 978-1-78945-075-0

ERC code:
PE2 Fundamental Constituents of Matter
 PE2_13 Thermodynamics

Contents

Foreword xi
Michel FEIDT

Preface xiii
Bernard DESMET

Chapter 1. Energy Conversion: Thermodynamic Basics 1
Georges DESCOMBES and Bernard DESMET

    1.1. Introduction 1
    1.2. Principles of thermodynamics 2
        1.2.1. Notion of a thermodynamic system 2
        1.2.2. First law 2
        1.2.3. Second law: mechanism of mechanical energy degradation
        in a heat engine 5
    1.3. Thermodynamics of gases 14
        1.3.1. Equations of state 14
        1.3.2. Calorimetric coefficients 15
        1.3.3. Ideal gas 16
        1.3.4. Van der Waals gas 20
    1.4. Conclusion 21
    1.5. References 21

Chapter 2. Internal Combustion Engines 23
Bernard DESMET

    2.1. Generalities – Operating principles 23
        2.1.1. Introduction 23
        2.1.2. Spark-ignition engines 25
        2.1.3. Compression ignition engine 28
        2.1.4. Expression of useful work 29
    2.2. Theoretical air cycles 30
        2.2.1. Hypotheses 30
        2.2.2. Beau de Rochas cycle (Otto cycle) 31
        2.2.3. Miller–Atkinson cycle 36
        2.2.4. Diesel cycle 39
        2.2.5. The limited pressure cycle (mixed cycle) 41
        2.2.6. Comparison of theoretical air cycles 43
    2.3. Influences of the thermophysical properties of the working fluid
    on the theoretical cycles 43
        2.3.1. Thermophysical properties of the working fluid 43
        2.3.2. Reversible adiabatic transformations 45
        2.3.3. Mixed cycle for ideal and semi-ideal gases 47
    2.4. Zero-dimensional thermodynamic models 51
        2.4.1. Hypotheses 51
        2.4.2. Single-zone model 52
        2.4.3. Flow through the valves 54
        2.4.4. Heat transfer with the cylinder walls 56
        2.4.5. Combustion heat generation model 57
        2.4.6. Two-zone model 59
    2.5. Supercharging of internal combustion engines 61
        2.5.1. Basic principles of supercharging 61
        2.5.2. Supercharging by a driven compressor 62
        2.5.3. Turbocharging 63
    2.6. Conclusions and perspectives 66
    2.7. References 67

Chapter 3. Aeronautical and Space Propulsion 69
Yannick MULLER and François COTTIER

    3.1. History and development of aeronautical means of propulsion 69
    3.2. Presentation of the aircraft system and its propulsive unit 72
        3.2.1. Classification and presentation of the usual architectures
        of aeronautical engines and their specific uses 72
        3.2.2. Study of the forces applied on the aircraft system during
        steady flight 79

3.2.3. Definition of the propulsion forces and specific quantities
of the propulsion system 83
3.3. Operating cycle analysis 86
   3.3.1. Hypotheses and limits of validity 86
   3.3.2. Presentation of engine stations (SAE ARP 755 STANDARD) ... 88
   3.3.3. Study of thermodynamic transformations and their
   representations in T – s diagrams 91
   3.3.4. Study of the thermodynamic cycles for a gas turbine 93
   3.3.5. Study of the thermodynamic cycle of a gas turbine,
   branch by branch 96
   3.3.6. Improvements to the Joule–Brayton cycle 98
   3.3.7. Thermodynamic improvements for a gas turbine using energy
   regeneration 101
   3.3.8. Thermodynamic improvements for a gas turbine using staged
   compression and expansion 103
3.4. The actual engine 104
   3.4.1. Development cycle of the turbomachine (turbojet) 104
   3.4.2. Technical disciplines in development 108
   3.4.3. Some specific problems of each module 111
   3.4.4. Secondary air system design methods 125
   3.4.5. T_4 and the secondary air system 126
3.5. Perspectives 132
3.6. References 132

Chapter 4. Combustion and Conversion of Energy 133
Bernard DESMET

4.1. Generalities 133
   4.1.1. Introduction 133
   4.1.2. Premixed flame 135
   4.1.3. Diffusion flame 136
   4.1.4. Stabilization of a flame 137
   4.1.5. Flammability of air–fuel mixtures 138
   4.1.6. Combustion in internal combustion engines 139
4.2. Theoretical combustion reactions 141
   4.2.1. Constituents of the combustible mixture 141
   4.2.2. Combustion stoichiometry 142
   4.2.3. Theoretical combustion of a lean mixture 143
   4.2.4. Theoretical combustion of a rich mixture 144
4.3. Energy study of combustion 144
   4.3.1. Combustion at constant volume 144
   4.3.2. Combustion at constant pressure 146
   4.3.3. Relations between heating values 147

4.3.4. Adiabatic flame and explosion temperatures 150
4.4. Chemical kinetics of combustion 155
   4.4.1. Chain reactions 155
   4.4.2. Composition of a reactive mixture 156
   4.4.3. Reaction rates 157
   4.4.4. Establishing a chemical equilibrium 159
   4.4.5. Equilibrium composition of the combustion products 160
   4.4.6. Detailed chemical kinetics–formation of pollutants 164
4.5. Exergy analysis of combustion 166
   4.5.1. Exergy of a gas mixture 166
   4.5.2. Exergy production from a combustion reaction 169
   4.5.3. Exergy of a fuel 172
4.6. Conclusion 176
4.7. References 176

Chapter 5. Engines with an External Heat Supply 179
Georges DESCOMBES and Bernard DESMET

5.1. Introduction 179
5.2. The Stirling engine 180
   5.2.1. Theoretical cycle 180
   5.2.2. Characteristics of the Stirling engine 184
5.3. The Ericsson engine 187
   5.3.1. Operating principles 187
   5.3.2. Theoretical cycles 188
   5.3.3. Improvements of the Ericsson engine 193
5.4. Perspectives 194
   5.4.1. Advantages and disadvantages of Stirling and Ericsson engines . . 194
   5.4.2. Perspectives of evolution of external combustion machines
   in the new decarbonized energy landscape 195
5.5. References 195

Chapter 6. Energy Recovery – Waste Heat Recovery 197
Mohamed MEBARKIA

6.1. Waste energy recovery 197
   6.1.1. Energy balance of an internal combustion engine 197
   6.1.2. Degradation of mechanizable energy into uncompensated heat . . . 200
   6.1.3. Exergy balance in internal combustion engines 203
   6.1.4. Concept of energy recovery 205
6.2. Cogeneration in industrial facilities 205
   6.2.1. Cogenerating gas turbines 205
   6.2.2. Cogenerating diesel engine 207

6.2.3. Comparative cogeneration efficiencies 209
6.2.4. Complex depressurized cycle 211
6.2.5. Complex over-expansion cycle 212
6.2.6. Conclusion 214
6.3. Micro-cogeneration 215
  6.3.1. Introduction 215
  6.3.2. Classification 216
  6.3.3. Internal combustion engines 217
  6.3.4. Gas micro-turbines 218
  6.3.5. Fuel cells 221
  6.3.6. Thermoelectricity 223
  6.3.7. Thermoacoustics 224
  6.3.8. "Rankinized" cycles 225
6.4. Conclusion 227
6.5. Perspectives 228
6.6. References 229

List of Authors 233

Index 235

Foreword

Michel FEIDT

LEMTA, Université de Lorraine, Vandoeuvre-les-Nancy, France

The book you have in your hands is one of the books from the "Physics of Energy and Energy Efficiency" subject in the Engineering and Systems department.

The subject of "Physics of Energy and Energy Efficiency", albeit recent, is not new. It is particularly underpinned by a thermodynamic approach, whatever the scale.

The selected aspect will be phenomenological and characterized explicitly in order to emphasize the key concept of "efficiency", essential for any system or process.

The characterization chosen for the development of this subject has been arranged into four successive books, each strongly correlated with each other, and also with other series within the department:

– *Fundamental Physics of Energy*;

– *Thermodynamics of Heat Engines*;

– *Heat Engines with Inverse Cycles*;

– *Efficiency in Practice*.

I would like to thank ISTE, the various coordinators, and the authors for their contributions and effective actions, despite the very particular conditions of the moment. We are awaiting and would like to encourage comments, suggestions and questions from readers.

Thermodynamics of Heat Engines,
coordinated by Bernard DESMET.
© ISTE Ltd 2022.

Preface

Bernard DESMET

INSA – HdF, Université Polytechnique Hauts-de-France, Valenciennes, France

This book aims to present a synthesis of the basics of thermodynamics to understand and analyze the conversion of heat into work in heat engines. In particular, it is aimed at students with bachelor's and master's degrees in science and engineering, and it may also be useful for engineers and technicians working in the respective fields.

Within the fields of transport and industrial applications, the conversion of heat into mechanical work faces two major challenges: the preservation of natural resources, which implies a high energy efficiency, and the limitation of waste carbon dioxide and emission of pollutants. To measure the overall impact of mechanical energy production, it is necessary to distinguish between primary (coal, oil, natural gas, nuclear, etc.) and renewable (hydraulic, biomass, geothermal, solar, wind power, etc.) energy sources and carriers, i.e. energy resources which are not (or only slightly) available in their original state and which must be processed (electrical energy, hydrogen, gas if it is refined, etc.). The usefulness of energy carriers – which are intermediaries between primary energy sources and their destinations – lies in their capacity to store energy, their ability to be distributed, the resolution of local pollution problems as in the case of cities, and probably others. Their impact on global CO_2 emissions, and therefore on the climate, is linked to their production process. From a resource consumption point of view, the whole chain, from the source to its destination, should be examined.

For many years to come, it is very likely that the conversion of heat, from combustion or other sources, into mechanical work will remain necessary. This conversion process must therefore be optimized to preserve the primary energy resources and control its environmental impact. High-performance energy conversion systems, based on combined cycles or cogeneration installations, mostly use internal combustion engines or gas turbines, whose performance must be equally optimized. The thermodynamic analysis developed in this work is one of the answers to these objectives.

The chapters of this book are as follows:

Chapter 1, *"Energy Conversion: Thermodynamic Basics"*, briefly presents the main thermodynamic concepts which are used in the rest of this book.

Chapter 2, *"Internal Combustion Engines"*, compares different theoretical operating cycles. Particular attention is focused on the thermodynamic properties of the working fluid. The more realistic zero-dimensional approach is then presented.

Chapter 3, *"Aeronautical and Space Propulsion"*, is devoted to the thermodynamic analysis of gas turbines and their applications in aeronautical propulsion. The authors present the different stages in the industrial process of developing turbojet engines.

Chapter 4, *"Combustion and Conversion of Energy"*, focuses on the energy aspects of combustion, which is the main source of heat for thermal machines.

Chapters 5 and 6, *"Engines with an External Heat Supply"* and *"Energy Recovery – Waste Heat Recovery"*, are motivated by the need for environmental protection. Motors with an external heat input, which are not widely used, are finding renewed interest due to their ability to use various heat sources, and the improvements made possible by technical progress. Chapter 6 deals with energy recovery, particularly in the form of high-power cogeneration or micro-cogeneration.

This book is the product of a close collaboration between the authors:

François Cottier is a graduate engineer from ENSIMEV, a school within the University of Valenciennes. He has 20 years of industrial experience in aeronautical propulsion at MTU Aero Engines in Munich. As a specialist in thermal analysis of turbojet components, he was responsible for the thermal design of turbine blades for several turbojets on the market, and leads many research projects for the development of new turbojet technologies. A specialist in numerical simulation of flows and heat transfers, he was responsible for the integration of modern 3D numerical simulation methods in fluid mechanics in the development process of turbojet engines. Currently, he participates in the development of future technologies and original concepts for aeronautical applications.

Georges Descombes is a CNAM engineer who served as a university professor at the CNAM in Paris from 2007 to 2018 after starting his university career as an assistant in 1990. He had previously experienced his apprenticeships and his job as a technician, then as an engineer in the private sector for 20 years in the automotive and industrial transport sectors. He has supervised around 100 engineering and university master's dissertations as well as around 20 doctoral theses. He has published around a 100 articles, international conferences in addition to numerous academic works. He is currently a Scientific Advisor in Energy Physics of Mobility.

Bernard Desmet is an engineer who graduated in 1970 from the Ecole Nationale Supérieure d'Arts et Métiers (ENSAM), worked as a research fellow at the University of Sciences and Techniques in Lille and at the ENSAM Lille, and then was a professor at the National School of Computer Science, Automation, Mechanics and Energetics and Electronics (ENSIAME) at the University of Valenciennes and Hainaut-Cambrésis from 1988 to 2011. He has supervised 27 doctoral theses. His research focuses on the transfer phenomena in fluid engines, with the aim of improving their performance with respect to energy consumption and reducing emissions. He is the author of more than 40 articles in various scientific journals.

Mohamed Mebarkia is an engineer who graduated in 2014 from the Faculty of Science and Technology of the University of Tébessa, specializing in electromechanics. He defended his doctoral thesis in 2018 at the Institut des Mines at the University of Tébessa, focusing on the study of thermal transport and the boiling mechanism with different configurations of heating tubes. This theme concerns thermal energy and the recovery of waste energy in energy production units in building, industry, heat engines, gas turbines and combined cycles, air conditioning and climate engineering, with low heat rejection in addition to second-generation renewable energies, and hot and cold cogeneration.

Yannick Muller is an engineer who graduated from the National School of Aeronautical Construction Engineers (ENSICA – Toulouse) in 2003. He defended his doctoral thesis at the University of Valenciennes and Hainaut-Cambrésis, which was prepared as part of a collaboration with the turbojet manufacturer MTU Aero Engines AG in Munich, and has been working in this company ever since. He has more than 15 years of experience in engine and propulsion technology development, holding various positions in the secondary air system (SAS) and Engine Performance teams, and simultaneously carried out several design method improvement projects to improve engine design methods in aeronautics. In charge of the SAS internal software at MTU, he promoted the use of coupled "fluid–structure–thermomechanical" modeling. He also introduced the use of stochastic methods for the SAS. After this position within the engine performance department, he now works on future engine concepts with the thermodynamic team.

Thanks are addressed to *Céline Morin*, professor at INSA Hauts-de-France/UPHF, for her careful proofreading of Chapter 4 as well as for her suggestions.

July 2022

1

Energy Conversion: Thermodynamic Basics

Georges DESCOMBES[1] and Bernard DESMET[2]

[1] *CNAM, Paris, France*
[2] *INSA – HdF, Université Polytechnique Hauts-de-France, Valenciennes, France*

1.1. Introduction

We are interested here in the *conversion of heat into mechanical work* via machines using a fluid medium in a continuous flow, or functioning in a cyclic manner. This first chapter succinctly presents the main concepts of thermodynamic used in this context. For a more in-depth study, the reader may consult the specialized works of Borgnakke and Sonntag (2013), Feidt (2014), Foussard et al. (2021) and Çengel et al. (2019).

Classical sign conventions will be used: the quantities of heat and work exchanged between a system and its exterior will be positive while they are received by the system.

Work, quantities of heat and extensive state quantities – quantities of which the value is proportional to the quantity of matter of the system – will be denoted in uppercase when they refer to the whole system and in lowercase when they are expressed per unit mass. Therefore, W, Q, U, etc. refer to work exchange, heat, internal energy, etc. for the considered system, and w, q, u, etc. are the corresponding specific quantities.

Thermodynamics of Heat Engines,
coordinated by Bernard DESMET.
© ISTE Ltd 2022.

1.2. Principles of thermodynamics

1.2.1. *Notion of a thermodynamic system*

In the strict sense of the term, a *thermodynamic system* or even a *closed system* does not exchange matter with its exterior. Its boundary is impermeable to the exchange of matter. In technical thermodynamics, we are interested more often in the equipment (heat exchangers, turbines, compressors, etc.) through which one or more fluids flow. Generally, we look for characteristics (pressure, temperature, mass flow rate, etc.) in the fixed sections located on either side of the component being studied and defined as the inlet and output of this component. These sections are continuously crossed by the flowing fluids. A closed control surface that comprises the inlet and outlet sections of the component therefore does not determine a closed system. This is called an *open system*. In the case of an open system, the material contained within the boundaries of the control surface is constantly renewed.

1.2.2. *First law*

For a closed system that evolves following a *cyclic transformation* (the final state coinciding with the initial state) by exchanging work W_e and heat Q_e with the exterior (Figure 1.1), the first law of thermodynamics expresses the equivalence between heat and work:

$$W_e + Q_e = 0 \qquad [1.1]$$

For a closed system that evolves from an initial state i into a final state f:

$$W_e + Q_e = U_f - U_i = \Delta U_{i-f} \qquad [1.2]$$

where U represents the internal energy of the considered system. U [J] is an *extensive state quantity*; therefore, it only depends on the state of the system.

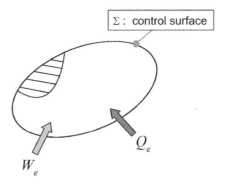

Figure 1.1. *Closed system: first law*

In the case when kinetic energy of the system plays a significant role, we can write:

$$W_e + Q_e = \Delta U_{i-f} + \Delta Ec_{i-f} \qquad [1.3]$$

where ΔEc_{i-f} represents the change in the kinetic energy of the system between the initial and final states.

Now let us consider the case of a machine that functions in a *steady state* through which a flowing fluid of mass flow rate q_m passes (Figure 1.2), comprising only a single inlet *in* and a single outlet *out* situated at the heights z_{in} and z_{out}, respectively.

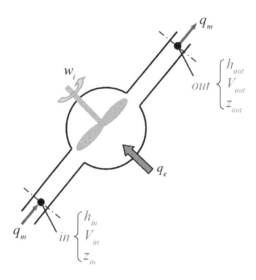

Figure 1.2. Open system with one inlet and one outlet in a steady state: first law

w_i [J/kg] is the indicated specific work, i.e. the mechanical work exchanged between the fluid and the moving parts of the machine (piston, rotor, etc.), and q_e is the quantity of heat per unit mass exchanged with the exterior of the fluid passing through the machine. h [J/kg] and V [m/s] are, respectively, the specific enthalpy and the velocity of the fluid flow:

$$h = u + p\,v \qquad [1.4]$$

where u [J/kg] is the specific internal energy, p [Pa] is the pressure, $v = 1/\rho$ [m³/kg] is the specific volume and ρ [kg/m³] is the density.

In this case, the first law of thermodynamics can be written as:

$$w_i + q_e = h_{out} - h_{in} + \frac{V_{out}^2 - V_{in}^2}{2} + g\,(z_{out} - z_{in}) \qquad [1.5]$$

where g is the gravitational acceleration on Earth.

Equation [1.5] can be generalized (Figure 1.3) to the case of a machine with several inlets and outlets. It is therefore necessary to take into account the relationships between the different flows in the inlet and outlet sections. The simplest way is to reduce the different quantities to the unit of time by introducing the power indicated (P_i [W]: the mechanical power exchanged between the fluids and the mobile parts of the machine) and the thermal flux exchanged with the exterior (Φ_e [W]: the quantity of heat exchanged with the exterior per unit time).

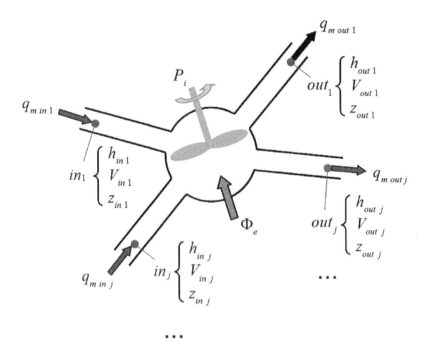

Figure 1.3. *Open system with several inlets and outlets in a steady state: first law*

The first law of thermodynamics can be written in the form:

$$P_i + \Phi_e = \sum_{j=1}^{n_{out}} q_{m\ out\ j} \left(h_{out\ j} + \frac{V_{out\ j}^2}{2} + g\, z_{out\ j} \right)$$
$$- \sum_{j=1}^{n_{in}} q_{m\ in\ j} \left(h_{in\ j} + \frac{V_{in\ j}^2}{2} + g\, z_{in\ j} \right) \qquad [1.6]$$

where n_{in} and n_{out} correspond to the number of inlets and outlets of the machine situated at the heights z_{in} and z_{out}, respectively. q_{min}, q_{mout}, h_{in}, h_{out}, V_{in}, V_{out} are the mass flow rates, specific enthalpies and fluid flow velocities, respectively, in the sections defined as inlets and outlets.

1.2.3. *Second law: mechanism of mechanical energy degradation in a heat engine*

1.2.3.1. *Concept of an engine operating between hot and cold heat energy reservoirs*

The transfer of thermal energy through a diathermal wall follows from the spontaneous transfer of the kinetic energy from the molecules of the hot face to the molecules of the cold face. The wall thus receives and yields the heat quantities transmitted, and this transfer of transport is none other than thermal diffusion via *conduction* (Figure 1.4).

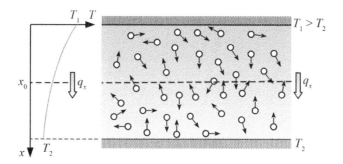

Figure 1.4. *Microscopic diffusion of energy*

This kinetic energy transfer tends towards an equilibrium state while the molecular collisions tend themselves towards a homogeneous partition of average kinetic energy, by using the molecules in the wall as an intermediary. A material body that tends to cool can even give rise to an electromagnetic field, which carries the equivalent of the molecular kinetic energy given up by the body as *radiation*.

The spontaneous movement of molecular kinetic energy therefore takes place from the hot points towards the cold points, and the conversion of the thermal energy into the kinetic energy hence can only be realized in a heat engine if it is placed between two distinct temperature levels. A fraction of the thermal energy is inevitably received by the receiving thermostat because if it did not receive any, the spontaneous flow would no longer exist. Hence, we return to the case of a single source for which we know that it is impossible to obtain any mechanical work. A *monothermal cycle* corresponds effectively to the trivial energy conversion, which is realized by a system exclusively subjected to friction.

It is therefore indispensable that a heat engine is placed between a source and a sink at different temperature levels to obtain a conversion of heat into work, and this is translated to the macroscopic level by *Carnot's concept*. This indicates that an *unavoidable energy* is inevitably rejected to the outside in the form of heat, even in the hypothesis of an ideal reversible cycle which is clearly not possible. The many irreversibilities inherent in the product of the real mechanical energy in an engine increase the energy degradation in a very significant way, hence the advantage of providing devices for recovering the unavoidable energy.

1.2.3.2. *Carnot's concept*

It was the eminent scientist Sadi Carnot (1796–1832) who had established in 1824 the fundamental concepts of macroscopic thermodynamics, which are applicable to the conversion of thermal energy into mechanical in the shaft of a motor. Carnot investigated the factors that govern the conversion of heat energy into mechanical energy for a steam machine, and for this he stated the concept of reversibility based on the following two considerations:

> Any mechanical irreversibility has the effect of reducing the mechanical work that can be collected from a machine operating according to a determined cycle which receives a quantity of heat from the heat source. Any difference in temperature between the elastic fluid and the sources of heat or within an elastic fluid is also the cause of a reduction in the quantity of mechanical work that a thermal machine is capable of producing from a given quantity of heat.

Carnot's external combustion engine (Figure 1.5) is placed between a source and a sink, and functions according to a theoretical cycle that is not the base of any aero-thermochemical irreversibility.

The fluid of work constitutes a *closed thermodynamic system* subjected to the following transformations:

– 1–2: the isothermal expansion at the temperature T_h during which the working fluid receives from the heat source the amount of heat Q_h;

– 2–3: the reversible adiabatic expansion from the temperature T_h to the temperature T_c;

– 3–4: the isothermal compression at the temperature T_c during which the working fluid transfers to the heat sink the amount of heat $|Q_c|$;

– 4–1: the reversible adiabatic compression during which the temperature of the working fluid passes from T_c to T_h.

According to the first law, the work done by the cycle is given by:

$$W + Q_h + Q_c = 0 \qquad [1.7]$$

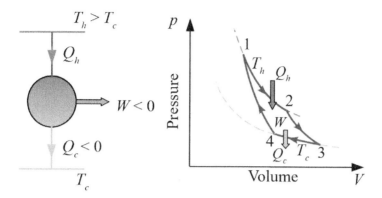

Figure 1.5. *Carnot's external combustion engine*

The thermodynamic efficiency of Carnot's engine cycle, in the sense of the first law, is defined by the relation:

$$\eta_c = \frac{|W|}{Q_h} = 1 - \frac{T_c}{T_h} \qquad [1.8]$$

The efficiency of such a cycle constitutes an upper bound.

By using relations [1.7] and [1.8]:

$$-W = \eta_c\, Q_h = \left(1 - \frac{T_c}{T_h}\right) Q_h = Q_h + Q_c \qquad [1.9]$$

we deduce:

$$\frac{Q_h}{T_h} + \frac{Q_c}{T_c} = 0 \qquad [1.10]$$

More generally, the variation of *entropy* S for a closed system during a *reversible* elementary transformation is:

$$d S_{rev} = \frac{\delta Q_e}{T} \qquad [1.11]$$

where δQ_e represents the quantity of elementary heat exchanged with the exterior, and T is the thermodynamic temperature of the interface where this exchange takes place. Entropy is an *extensive function of state*.

In the case of a real transformation and hence *irreversible*:

$$d S = d S_{rev} + d S_{irr} \qquad [1.12]$$

where dS_{irr} is *strictly positive* in the case of an irreversible transformation, and zero if the transformation is reversible.

According to equation [1.11], a *reversible adiabatic* transformation is also isentropic. The cycle of Carnot's combustion engine is here composed of two isotherms and two isentropes (Figure 1.6).

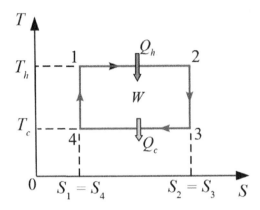

Figure 1.6. *Carnot's external combustion engine cycle*

1.2.3.3. *Carnot's paradox*

The concept of Carnot's reversible cycle imposes the hypothesis of reversibility on the thermal exchanges occurring at the hot source and the cold sink. This reversibility condition, which is based on the concept of steady-state thermodynamics, assumes that the thermal gradients between the hot source, the elastic fluid and the cold sink are all zero.

This hypothesis is obviously not possible, in contradiction with the laws of heat transmission, since the existence of a temperature gradient is mandatory for the system to be the base of a heat flow.

The first law of thermodynamics characterized by Fourier's phenomenological law (equation [1.13]) describes the case of heat transfer via thermal conduction in an isotropic material, along the one-dimensional Cartesian coordinate x:

$$q_x = -\lambda \frac{\partial T}{\partial x} \qquad [1.13]$$

where q_x is the heat flux density expressed in W/m^2, and λ is the heat conductivity expressed in W/(m · K). The negative sign is imposed by the second law of thermodynamics, and indicates that the heat transfer takes place from the hot side to the cold side.

In the case of a heat source of uniform temperature T_1, which generates the heat flux $\delta Q/dt$ to an environment of uniform temperature T_2 ($T_1 > T_2$), the entropy flow resulting from the irreversibility of the transfer is given by:

$$\frac{dS_{irr}}{dt} = \left(\frac{1}{T_2} - \frac{1}{T_1}\right)\frac{\delta Q}{dt} \qquad [1.14]$$

To edge towards the condition of reversibility, we can imagine a tiny difference between the temperatures T_1 and T_2. In accordance with equation [1.13], this assumption would nevertheless lead to a heat flux density q_x that tends towards zero, and the motive power delivered by such a machine would also tend towards zero.

1.2.3.4. Endoreversible Carnot cycle

A cycle is said to be endoreversible if the internal exchanges are carried out according to reversible evolutions, but the external exchanges give rise to irreversibilities.

As indicated in the previous section, heat transfers between the hot source, working fluid and the heat sink require finite temperature differences which constitute sources of irreversibility. Thus, the amount of heat Q_h provided by the hot source of temperature T_h is used by the Carnot engine at a temperature $\overline{T}_h$ that is lower than T_h, and the amount of heat Q_c transferred by the working fluid at the temperature $\overline{T}_c$ to the heat sink at the temperature T_c is lower than $\overline{T}_c$ (Figure 1.7).

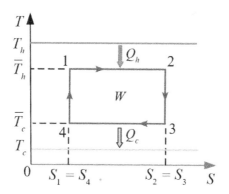

Figure 1.7. *An endoreversible dithermal cycle*

According to the second law of thermodynamics:

$$\frac{Q_h}{\overline{T}_h} + \frac{Q_c}{\overline{T}_c} + \Delta S_{irr} = 0 \qquad [1.15]$$

where ΔS_{irr} is the increase in entropy due to temperature differences. By expressing Q_c:

$$-Q_c = Q_h \frac{\overline{T}_c}{\overline{T}_h} + \overline{T}_c \Delta S_{irr} \qquad [1.16]$$

and by transforming [1.16] into the expression of the first law of thermodynamics (equation [1.7]):

$$W = \overline{T}_c \Delta S_{irr} - Q_h \left(1 - \frac{\overline{T}_c}{\overline{T}_h}\right) \qquad [1.17]$$

we deduce the thermodynamic efficiency of the endoreversible cycle to be:

$$\eta_{th} = \frac{|W|}{Q_h} = \left(1 - \frac{\overline{T}_c}{\overline{T}_h}\right)\left(1 - \frac{\overline{T}_c \Delta S_{irr}}{Q_h\left(1 - \frac{\overline{T}_c}{\overline{T}_h}\right)}\right) \qquad [1.18]$$

REMARK.– If $\Delta S_{irr} = 0$ (reversible cycle), then:

$$\eta_{th} = 1 - \frac{\overline{T}_c}{\overline{T}_h} = \eta_c \qquad [1.19]$$

which is the endoreversible efficiency of the Carnot cycle.

1.2.3.5. *Exergy of a closed system*

In the case of a closed system that evolves from an initial state i to a final state f, for example under the action of a piston as represented in Figure 1.8, by taking into account the mechanical equilibrium equation for the piston, the work exchanged with the volume V of the elastic fluid is given by:

$$W_e = \int_{(i \to f)} -p\, dV = W_m + \int_{(i \to f)} -p_a\, dV \qquad [1.20]$$

The piston is assumed to be in equilibrium under the action of the fluid pressure p contained in the cylinder, the pressure p_a of the environment and the force F of the mechanical actions applied to the piston. The displacement force F of the piston corresponds to the useful mechanical work W_m that is exchanged with the exterior.

According to the first law of thermodynamics for an elementary transformation:

$$\delta W_m - p_a\, dV + \delta Q_e = dU \qquad [1.21]$$

and according to the second law:

$$dS = \frac{\delta Q_e}{T} + dS_{irr} \qquad [1.22]$$

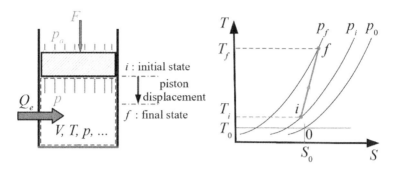

Figure 1.8. *Transformation of a closed system*

dU and dS are the internal energy and the entropy of the fluid contained in the cylinder, respectively, and $dS_{irr} > 0$ is the increase in elementary entropy associated with the irreversibility of the transformation.

Using equations [1.21] and [1.22]:

$$d(U - T_0 S) = \delta W_m - p_a\, dV + \delta Q_e - \frac{T_0}{T} \delta Q_e - T_0\, dS_{irr} \qquad [1.23]$$

$$\delta W_m + \left(1 - \frac{T_0}{T}\right) \delta Q_e = d(U - T_0 S + p_a V) + T_0\, dS_{irr} \qquad [1.24]$$

The index "0" represents a *reference state* that may be chosen arbitrarily.

By integrating equation [1.24] from the initial state i to the final state f of the transformation:

$$W_m + \int_{(i \to f)} \left(1 - \frac{T_0}{T}\right) \delta Q_e = \\ U_f - U_i - T_0 (S_f - S_i) + p_a (V_f - V_i) + T_0 \Delta S_{irr} \qquad [1.25]$$

In the case of a reversible transformation between any state (index "j") and the reference state:

$$W_m + \int_{(j \to 0)} \left(1 - \frac{T_0}{T}\right) \delta Q_e = \\ U_j - U_0 - T_0 (S_j - S_0) + p_a (V_j - V_0) \qquad [1.26]$$

and the following relation defines the exergy of the closed system, considered in the state "j":

$$Ex_j = U_j - U_0 - T_0 (S_j - S_0) + p_a (V_j - V_0) \qquad [1.27]$$

According to equation [1.26], the exergy of a closed system in a given state corresponds to the sum of the useful mechanical work and the part of the quantities of heat exchanged with the outside that is likely to be converted into mechanical work in Carnot machines by bringing the system back to the reference state 0. The expression of exergy could be completed further by including the kinetic energy and potential energy terms of the situation (action of the forces of gravity).

Now that we have introduced exergy, expression [1.25] can be written as:

$$W_m + \int_{(i \to f)} \left(1 - \frac{T_0}{T}\right) \delta Q_e = Ex_f - Ex_i + Ex_d \qquad [1.28]$$

where:

$$Ex_d = \int_{(i \to f)} T_0 \, dS_{irr} \geq 0 \qquad [1.29]$$

Ex_d is the destroyed exergy, a strictly positive quantity in the case of a real transformation, and hence irreversible.

1.2.3.6. *Exergy of an open system*

To simplify the presentation, the case considered here is that of a machine with a single inlet and a single outlet (Figure 1.2), for which the first law of thermodynamics can be written in the form:

$$w_i + q_e = h_{out} - h_{in} + \frac{V_{out}^2 - V_{in}^2}{2} + g(z_{out} - z_{in}) \qquad [1.5]$$

Assuming that the heat exchange with the outside is continuously distributed between the inlet and the outlet of the machine, we then have according to the second law of thermodynamics:

$$s_{out} - s_{in} = \int_{(in \to out)} \frac{\delta q_e}{T} + \Delta s_{irr} \qquad [1.30]$$

where s_{in} and s_{out} are the specific entropies. Combining equations [1.5] and [1.30]:

$$w_i + \int_{(in \to out)} \left(1 - \frac{T_0}{T}\right) \delta q_e =$$
$$h_{out} - h_{in} - T_0(s_{out} - s_{in}) + \frac{V_{out}^2 - V_{in}^2}{2} + g(z_{out} - z_{in}) + T_0 \Delta s_{irr} \qquad [1.31]$$

By defining the specific exergy in the form:

$$ex = h - h_0 - T_0(s - s_0) + \frac{V^2 - V_0^2}{2} + g(z - z_0) \qquad [1.32]$$

equation [1.31] can be written as:

$$w_i + \int_{(in \to out)} \left(1 - \frac{T_0}{T}\right) \delta q_e = ex_{out} - ex_{in} + ex_d \qquad [1.33]$$

The term:

$$ex_d = T_0 \, \Delta s_{irr} \geq 0 \qquad [1.34]$$

is the destroyed specific exergy, a strictly positive quantity for an irreversible transformation.

The previous discussion can be generalized to the case of a machine with several inlets and outlets (Figure 1.3). The thermal flow Φ_e exchanged with the outside results from the transfer with n_k sources or sinks that supply or receive the flows Φ_{ek}, and whose exchange surface temperatures are T_k:

$$\Phi_e = \sum_{k=1}^{n_k} \Phi_{ek} \qquad [1.35]$$

and

$$P_i + \sum_{k=1}^{n_k} \left(1 - \frac{T_0}{T_k}\right) \Phi_{ek} = \sum_{j=1}^{n_{out}} q_{m\ out\ j}\ ex_{out\ j} \\ - \sum_{j=1}^{n_{in}} q_{m\ in\ j}\ ex_{in\ j} + \dot{E}x_d \qquad [1.36]$$

where the power indicated by P_i and the thermal fluxes Φ_{ek} are counted positively when supplied to the fluids circulating in the machine. The destroyed exergy term per unit time:

$$\dot{E}x_d = T_0 \, \frac{dS_{irr}}{dt} \geq 0 \qquad [1.37]$$

is strictly positive in the case of irreversible changes.

REMARKS.–

– The above-presented exergetic approach only takes physical exergy (mechanical and thermal) into account. In the case of systems whose compositions vary, it is necessary to complete the expression of exergy by including chemical exergy terms (Benelmir et al. 2002; Benelmir 2018).

– The expression of the exergy requires defining a reference state (index "0"). In admitting (Lallemand 2007),

> which is a realistic engineering point of view, that the cold source which we might freely access is the surrounding environment,

the reference state is that of the system when it is brought back into an equilibrium from the ambient environment.

1.3. Thermodynamics of gases

1.3.1. *Equations of state*

For a gas in *thermodynamic equilibrium*, pressure, temperature and volume are related by an equation, which can be written in an implicit form as:

$$f(p, V, T) = 0 \qquad [1.38]$$

For an *ideal gas*, the Boyle–Mariotte state equation, established in 1662, can be written as:

$$p\,V = R\,T \qquad [1.39]$$

where V is the molecular volume and R is the gas constant for ideal gases ($R = 8.314$ J/(mol · K)).

By dividing the two terms in the equation of state by the molecular mass M, this can be expressed as a function of the specific quantities:

$$p\,v = r\,T \qquad [1.40]$$

where $v = V/M$ is the specific volume and $r = R/M$ is a specific constant that depends on the considered gas. For example, with dry air, $r = 287$ J/(kg · K). The ideal gas approximation is generally sufficient at low pressures.

Several equations have been established to improve the description of the behavior for real gases. Among these equations, the one established by van der Waals in 1873 is:

$$\left(p + \frac{A}{V^2}\right)(V - B) = R\,T \qquad [1.41]$$

where A and B are constants and relative to the specific quantities, the van der Waals equation can be written as:

$$\left(p + \frac{a}{v^2}\right)(v - b) = r\,T \qquad \text{with: } a = \frac{A}{M^2} \text{ and } b = \frac{B}{M} \qquad [1.42]$$

For air, the following values can be used:

$A = 0.136$ Pa · m⁶/mol²	$B = 0.0362 \cdot 10^{-3}$ m³/mol
$a = 162.1$ Pa · m⁶/kg²	$b = 1.250 \cdot 10^{-3}$ m³/kg

Table 1.1. *Constants used in the van der Waals equation for air*

1.3.2. *Calorimetric coefficients*

The study of calorimetric coefficients is classical (Feidt 2014), and only a few results are recalled here. For a gas, two intensive variables are sufficient to characterize the state of a system, such as the temperature and the pressure, or the temperature and the specific volume. The heat quantities exchanged per unit mass of gas can be expressed in the form:

$$\delta q = c_v(T,v)\, dT + l_T(T,v)\, dv \qquad [1.43]$$

$$\delta q = c_p(T,p)\, dT + h_T(T,p)\, dp \qquad [1.44]$$

where c_v and c_p are the *specific heat capacities* at a constant volume and pressure, respectively.

The differential expressions for the specific values of the internal energy, enthalpy and entropy are:

$$du = \delta q - p\, dv = c_v\, dT + (l_T - p)\, dv \qquad [1.45]$$

$$dh = \delta q + v\, dp = c_p\, dT + (h_T + v)\, dp \qquad [1.46]$$

$$ds = \frac{c_v}{T}\, dT + \frac{l_T}{T}\, dv = \frac{c_p}{T}\, dT + \frac{h_T}{T}\, dp \qquad [1.47]$$

According to equations [1.45] and [1.46]:

$$c_v = \left(\frac{\partial u}{\partial T}\right)_v \qquad c_p = \left(\frac{\partial h}{\partial T}\right)_p \qquad [1.48]$$

The calorimetric coefficients l_T and h_T can be obtained by expressing du, dh and ds as the differentials of physical state quantities:

$$l_T = T\left(\frac{\partial p}{\partial T}\right)_v \qquad h_T = -T\left(\frac{\partial v}{\partial T}\right)_p \qquad [1.49]$$

as well as the generalized Mayer's relation:

$$c_p - c_v = T \left(\frac{\partial p}{\partial T}\right)_v \left(\frac{\partial v}{\partial T}\right)_p \quad [1.50]$$

1.3.3. *Ideal gas*

For a gas obeying the equation of state for ideal gases, equation [1.40], then according to relation [1.49], we have:

$$l_T = p \qquad h_T = -v \quad [1.51]$$

Following from this:

$$du = c_v \, dT \qquad dh = c_p \, dT \quad [1.52]$$

and, as du and dh are the differentials of physical state quantities, c_v and c_p must be functions solely of the temperature. The expression for the entropy differential reduces to:

$$ds = c_v \frac{dT}{T} + r \frac{dv}{v} = c_p \frac{dT}{T} - r \frac{dp}{p} \quad [1.53]$$

and Mayer's relation reduces to:

$$c_p(T) - c_v(T) = r \quad [1.54]$$

The strict definition of an ideal gas not only imposes it to obey the state equation [1.39], but also the constant value hypothesis for the thermal capacities c_v and c_p. Therefore, in setting:

$$\gamma = \frac{c_p}{c_v} \quad [1.55]$$

and using Mayer's relation:

$$c_v = \frac{r}{\gamma - 1} \qquad c_p = \frac{\gamma r}{\gamma - 1} \quad [1.56]$$

We can define a semi-ideal gas (Figure 1.9) as a gas that respects the state equation [1.39] but for which the thermal capacities depend on the temperature.

For a reversible adiabatic transformation, and hence isentropic, between the initial state i and the final state f, according to [1.53], we have:

$$\int_{(i \to f)} c_v(T) \frac{dT}{T} + r \ln\left(\frac{v_f}{v_i}\right) = 0 \quad [1.57]$$

$$\int_{(i \to f)} c_p(T) \frac{dT}{T} - r \ln\left(\frac{p_f}{p_i}\right) = 0 \qquad [1.58]$$

In the case of an ideal gas corresponding to the strict definition, equations [1.57] and [1.58] make it possible to find classical expressions for reversible adiabatic transformations:

$$p\,v^\gamma = C^{te} \qquad T\,v^{\gamma-1} = C^{te} \qquad \frac{T}{p^{\frac{\gamma-1}{\gamma}}} = C^{te} \qquad [1.59]$$

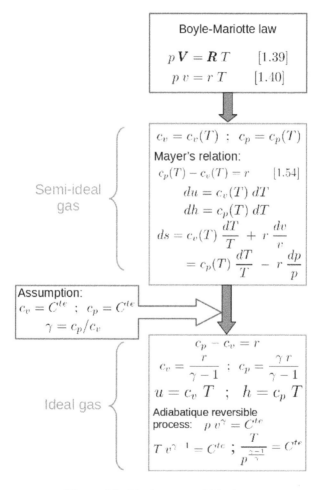

Figure 1.9. *Ideal gas–semi-ideal gas*

Heat capacities are usually modeled polynomially in many publications. In particular, the coefficients of the polynomials for the calculation of heat capacities at constant pressure for many species are given by McBride et al. (1933) in the form:

$$\frac{C_p^0}{R} = \frac{c_p^0}{r} = a_1 + a_2\,T + a_3\,T^2 + a_4\,T^3 + a_5\,T^4 \qquad [1.60]$$

where C_p^0 and c_p^0 are the molar and specific heat capacities at low pressures and a constant pressure, respectively. Table 1.2 gives the values of the coefficients for the polynomials taken from McBride et al. (1933) for some chemical species. To obtain a good accuracy of the values, different polynomials are defined, respectively, for the low temperature range (200–1,000 K) and the high temperature range (1,000–6,000 K).

Species	a_1	$a_2 \times 10^3$	$a_3 \times 10^6$	$a_4 \times 10^9$	$a_5 \times 10^{12}$	$T_{min} - T_{max}$ [K]
Nitrogen N_2	3.53101	-0.12366	-0.50300	2.43531	-1.40881	200–1 000
	2.95258	1.39690	-0.49263	0.078601	-0.00461	1 000–6 000
Oxygen O_2	3.78246	-2.99673	9.8473	-9.6813	3.24373	200–1 000
	3.66096	0.656366	-0.14115	0.0205798	-0.00130	1 000–6 000
Carbon dioxide CO_2	2.35677	8.9846	-7.12356	2.45919	-0.1437	200–1 000
	4.63659	2.74132	-0.99583	0.160373	-0.00916	1 000–6 000
Carbon monoxide CO	3.57953	-0.61035	1.01681	0.90701	-0.90442	200–1 000
	3.04849	1.35173	-0.48579	0.0788537	-0.00470	1 000 - 6 000
Water (vapor) H_2O	4.19864	-2.0364	6.5204	-5.48797	1.77198	200 - 1 000
	2.67704	2.97318	-0.77377	0.0944337	-0.00427	1 000–6 000

Table 1.2. *Coefficients for the calculation of thermal capacities according to McBride et al. (1933)*

Figure 1.10 shows the variations of specific heat capacities at low pressure for N_2, O_2, CO_2, CO and for water vapor in the temperature range from 200 K to 2,500 K obtained by using the polynomial coefficients from the work of McBride et al. (1933). It is clear that the heat capacities cannot be considered as constants when the temperature variations are significant, in order to obtain an acceptable accuracy of the calculation results.

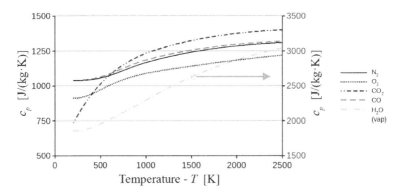

Figure 1.10. *Specific heat capacities at constant pressure as a function of temperature*

In a mixture in which each gas satisfies the ideal gas equation, the proportions of the different species determined by their volume are equal to their molar proportions. The molar heat capacity at constant pressure is given by:

$$C_p^0 = \sum_i \left(\frac{n_i}{n}\right) C_p^0(A_i) \qquad [1.61]$$

where $C_p^0(A_i)$ is the molecular heat capacity at a constant pressure for the species A_i, and n_i and n are the number of moles of the species A_i and the total number of gas molecules contained in the mixture, respectively.

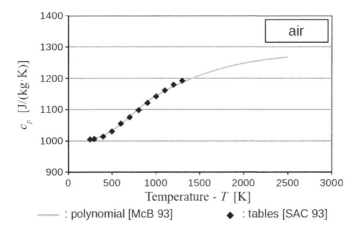

Figure 1.11. *Specific heat capacity at a constant air pressure as a function of temperature*

For energy applications, carbon dioxide and the rare gases contained in the air more often play a negligible role and air can be considered as a mixture of 20.9% oxygen and 79.1% nitrogen, corresponding to the chemical representation $O_2 + 3.76\ N_2$. Figure 1.11 shows a good agreement between the specific heat capacities of air as a function of temperature obtained using the polynomials from the work of McBride et al. (1933), and those appearing in the tables given by Saccadura (1993).

1.3.4. *Van der Waals gas*

For a van der Waals gas, its equations of state is:

$$\left(p + \frac{a}{v^2}\right)(v - b) = r\,T \qquad [1.42]$$

$$p = \frac{r\,T}{v - b} - \frac{a}{v^2} \quad \text{and}: \quad \left(\frac{\partial p}{\partial T}\right)_v = \frac{r}{v - b} \qquad [1.62]$$

According to equation [1.49]:

$$l_T = T\left(\frac{\partial p}{\partial T}\right)_v = \frac{r\,T}{v - b} = p + \frac{a}{v^2} \qquad [1.63]$$

and by substituting equation [1.63] into equations [1.45] and [1.47]:

$$du = c_v\,dT + a\,\frac{dv}{v^2} \qquad [1.64]$$

$$ds = c_v\,\frac{dT}{T} + r\,\frac{dv}{v - b} \qquad [1.65]$$

with du being the differential of a physical quantity of state:

$$\left(\frac{\partial c_v}{\partial v}\right)_T = \left(\frac{\partial \frac{a}{v^2}}{\partial T}\right)_v = 0 \qquad [1.66]$$

We deduce from this that c_v does not depend on the volume, and hence c_v is uniquely a function of the temperature:

$$c_v = c_v(T) \qquad [1.67]$$

By using the generalized Mayer's relation:

$$c_p - c_v = T\left(\frac{\partial p}{\partial T}\right)_v \left(\frac{\partial v}{\partial T}\right)_p = -T\,\frac{\left(\frac{\partial p}{\partial T}\right)_v^2}{\left(\frac{\partial p}{\partial v}\right)_T} \qquad [1.68]$$

we obtain:

$$c_p(T,v) = c_v(T) + \cfrac{r}{1 - \cfrac{2\,a}{r\,T}\,\cfrac{(v-b)^2}{v^3}} \qquad [1.69]$$

The calculation of the internal energy and entropy can be carried out by integrating relations [1.64] and [1.65] between any chosen reference state (index 0) and the state in consideration:

$$u - u_0 = \int_{T_0}^{T} c_v(T)\,dT - a\left(\frac{1}{v} - \frac{1}{v_0}\right) \qquad [1.70]$$

$$s - s_0 = \int_{T_0}^{T} c_v(T)\,\frac{dT}{T} - r\,\ln\left(\frac{v_0 - b}{v - b}\right) \qquad [1.71]$$

1.4. Conclusion

This chapter has made it possible to recall some fundamental notions required when we study systems that operate with internal combustion engines, which will be used in the remainder of this work.

Particular emphasis was placed on the distinction between the approach based solely on the *first law* of thermodynamics, for which heat and work are equivalent quantities, and that based on the *second law*, which makes it possible to compare the real (and therefore irreversible) transformations of a system with the ideal (and reversible) evolutions of this system.

The concept of *exergy*, defined by combining the first and second principles of thermodynamics, makes a distinction between the different forms of energy (mechanical, thermal, chemical, etc.) by estimating their respective theoretical abilities to be converted into mechanical work.

1.5. References

Benelmir, R. (2018). *Définition et utilité de l'exergie*. Journées thématiques sur l'exergie, Nancy, 22–23 November.

Benelmir, R., Lallemand, A., Feidt, M. (2002). *Analyse exergétique – Définitions*. Les Techniques de l'ingénieur, BE 8015, pp. 1–15, Paris.

Borgnakke, C. and Sonntag, R.E. (2013). *Fundamentals of Thermodynamics*, 8th edition. John Wiley & Sons, Hoboken.

Çengel, Y.A., Boles, M.A., Kanoğlu, M. (2019). *Thermodynamics – An Engineering Approach*, 9th edition. McGraw Hill Education, New York.

Feidt, M. (2014). *Génie énergétique – Du dimensionnement des composants au pilotage des systèmes*. Dunod, Paris.

Foussard, J.-N., Julien, E., Mathé, H., Debellefontaine, H. (2021). *Les bases de la thermodynamique*, 3rd edition, Dunod, Paris.

Lallemand, A. (2007). Énergie, exergie, économie, thermo-économie. *Journées internationales de thermique*, Albi, 28–30 August.

McBride, B.J., Gordon, S., Reno M.A. (1993). Coefficients for calculating thermodynamic and transport properties of individual species. NASA Technical Memorandum 4513, October.

Saccadura, J.-F. (1993). *Initiation aux transferts thermiques*. Technique et Documentation, Lavoisier, Paris.

2
Internal Combustion Engines

Bernard DESMET

INSA – HdF, Université Polytechnique Hauts-de-France, Valenciennes, France

2.1. Generalities – Operating principles

2.1.1. *Introduction*

In 1862, the French Beau de Rochas invented the four-stroke cycle (intake, compression, combustion–expansion, exhaust). With the appearance of fuels making it possible to exploit this cycle, the German Nicholaus Otto then invented the first internal compression combustion engine in 1876. In 1897, Rudolf Diesel invented the compression ignition engine.

Most internal combustion engines use the action of a gaseous medium on a piston moving in a cylinder to create motive power (Figure 2.1). The piston inside of the cylinder undergoes an alternating translational movement, which is transmitted by the connecting rod to the crankshaft, thus converting this movement into a rotational one of the engine shaft. The engines can operate according to the *two-stroke* cycle (engine cycle carried out over one revolution of the crankshaft) or the *four-stroke* cycle (engine cycle carried out over two revolutions of the crankshaft). The four-stroke cycle is most often used, and thus we will limit ourselves to this case in the rest of the book.

Thermodynamics of Heat Engines,
coordinated by Bernard DESMET.
© ISTE Ltd 2022.

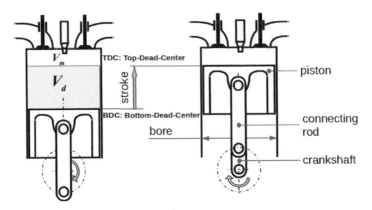

Figure 2.1. *Geometric characteristics*

The basic quantities used to characterize the geometry of an internal combustion engine are as follows:

– *displaced volume*, namely the volume swept by the displacement of the piston:

$$V_d = c\,\frac{\pi D^2}{4} \qquad [2.1]$$

where c is the stroke length of the piston, and D is the diameter of the cylinder (the bore). More generally, we may consider the engine cylinder's displacement, i.e. the product of the number of cylinders N_c in the engine with the displacement volume:

$$V_{d\_engine} = N_c\,c\,\frac{\pi D^2}{4} \qquad [2.2]$$

– *volumetric compression ratio*:

$$\tau = \frac{V_M}{V_m} \qquad [2.3]$$

where V_m is the *clearance volume*, namely, the residual volume at the bottom of the cylinder while the piston is at the top-dead-center (TDC) and $V_M = V_d + V_m$ is the volume of the cylinder while the piston is at the bottom-dead-center (BDC). The volumetric ratio is often incorrectly referred to as the "compression ratio".

Internal combustion engines can run on different fuels (combustibles) and, depending on the ignition mode, we will distinguish between:

– *spark-ignition* engines (gasoline, gas);
– *compression ignition* engines (diesel).

The power range of internal combustion engines is very large, and some orders of magnitude are listed in Table 2.1.

Applications	Scale model	Lawnmower	Automobile	Industrial heavy truck
Power [kW]	0.3–1.2	2–14	45–500	$150–20 \cdot 10^3$
Displacement volume [cm³]	2.5–15	50–500	1,000–2,200	$6 \cdot 10^3 – 4 \cdot 10^5$

Table 2.1. *Power ranges of different internal combustion engines*

2.1.2. *Spark-ignition engines*

We will not give detailed technological descriptions in this book, since these are available in several books (Fayette Taylor 1985; Heywood 1988; Pulkrabek 2020) or on the Internet.

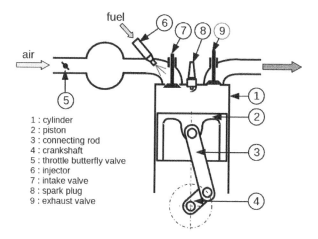

Figure 2.2. *Naturally aspirated spark ignition engine*

Figure 2.2 represents the main components of a naturally aspirated spark ignition engine (non-supercharged). Air intake occurs under ambient conditions (pressure p_a, temperature T_a). The butterfly valve placed within the inlet port system controls the mass of air that enters into the cylinder, which varies depending on the power required by the motor shaft. At a reduced power (partial load), closing the butterfly valve causes a head loss and the pressure p_b downstream from the valve (towards the cylinder) is lower than the ambient pressure p_a, allowing us to control the mass of air taken in. Fuel is either injected upstream from the butterfly valve into inlet port (indirect injection), or directly into the cylinder (direct injection), allowing for a greater control of the air–fuel proportions when compared to carburetors used in older engines. The substantially homogeneous air–fuel mixture undergoes combustion as a closed system, initiated by an electric spark generated by a spark plug when the piston is near the top-dead-center. The burnt gases are then discharged (exhaust) at the ambient pressure. The opening and closing of the valves control the intake and discharge of the gases in the cylinder, and are controlled mechanically by the camshaft. For an engine operating on a four-stroke cycle (the operating cycle requires two revolutions of the engine shaft), the camshaft drives at a rotational speed half that of the engine shaft.

The diagram in Figure 2.3 indicates the *timings* for the opening and closing of the valves, which are limited by factors such as their acceleration and the inertial effects from the fluid columns in the intake and exhaust ports. This leads to the opening and closing of the valves to be shifted with respect to the theoretical TDC and BDC positions, as to optimize how the cylinder *fills with air*.

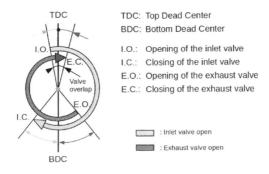

Figure 2.3. *Valve timing diagram*

The operating cycle of a four-stroke engine and the pressure–volume diagram are presented in Figures 2.4 and 2.5.

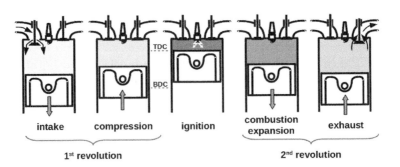

Figure 2.4. *Four-stroke operating cycle*

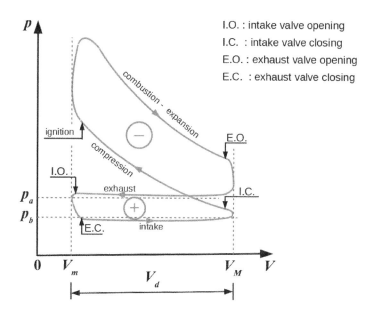

Figure 2.5. *Pressure–volume diagram*

The operating phases are as follows:

– *Intake*: the piston displaces itself from the TDC towards the BDC, and the inlet valve is open. The pressure in the cylinder is almost equal to the pressure p_b downstream from the butterfly valve, except for some pressure losses in the intake port and when passing by the valve. At full load, when the butterfly valve is fully open, there is essentially no pressure drop, $p_b \approx p_a$.

– *Compression*: the piston moves from the BDC to the TDC, and both valves are closed. The air–fuel mixture is formed upstream from the inlet valve in the case of indirect injection, or in the cylinder at the start of compression in the case of direct injection. There is an increase in temperature resulting from the compression which, if the compression ratio is too high, is likely to cause the self-ignition of the air–fuel mixture before that by the spark plug. This requires volumetric ratio to be limited (on the order of 10 for an automobile engine).

– *Combustion–expansion*: combustion is initiated by an electric spark produced by the spark plug, which is triggered pre-emptively just before the piston reaches the TDC, so that combustion occurs close to a constant volume evolution once at the TDC (ignition time, flame propagation time in the air–fuel mixture). The displacement of the piston from the TDC towards the BDC causes the hot gases to expand (in an isolated container) until the opening of the exhaust valve.

– *Exhaust*: the displacement of the piston from the BDC to the TDC ensures that the exhaust gases are discharged. Except for the pressure drops near the exhaust port, the pressure in the cylinder is equal to the ambient pressure p_a. The opening of the intake valve and the closing of the exhaust valve occur when the piston is close to the TDC.

2.1.3. *Compression ignition engine*

Figure 2.6 represents the schematics of the main components in a naturally aspirated (non-supercharged) compression ignition engine.

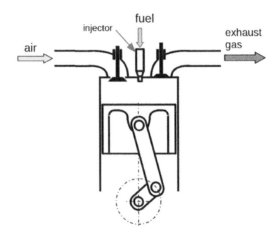

Figure 2.6. *Compression ignition engine*

Its four-stroke operating cycle is very similar to that of a spark-ignition engine; the way in which it differs from a spark-ignition engine is that there is no butterfly valve placed in the intake port, and the cylinder fills with *pure air* at a pressure close to the ambient pressure p_a. Therefore, the cylinder only contains air during the compression phase, and the fuel is introduced into the cylinder when combustion is about to occur. When fuel is introduced, the temperature in the cylinder must be high enough to cause it to *self-ignite*, which requires a sufficiently large compression ratio. Typically, the compression volume ratio for automotive engines is around the order of 20. Sufficient fuel atomization requires a fairly high *injection pressure*, compared to the pressure in the cylinder, which can reach 1,500 bar. The power delivered by the engine is adjusted by varying the quantity of fuel injected. The injectors can be supplied by either a high-pressure variable-flow pump, or via a ramp containing the fuel under pressure (common rail); in the latter case, the quantity of fuel injected is controlled by the opening time of the injectors. The expansion and exhaust phases are similar to those in the case of a spark-ignition engine.

2.1.4. *Expression of useful work*

Figure 2.7 represents the forces exerted on the gases evolving in the cylinder (a) and on the piston (b), respectively. The dynamic effects resulting from the accelerations of the moving elements (piston, connecting rods), in addition to the friction between piston and cylinder, are not considered here.

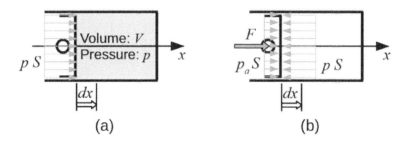

Figure 2.7. *Forces exerted on (a) the fluid and (b) the piston*

The elementary work transferred to the gas when the piston travels during the displacement dx of the piston is equal to:

$$\delta W = p\, S\, dx = -p\, dV \qquad [2.4]$$

where p is the pressure of the gas in the cylinder, S is the sectional area of the cylinder, $dV = -S\, dx$ is the variation of the piston cavity volume–cylinder, corresponding to the displacement dx.

For the entire operating cycle, the work done on the fluid is given by:

$$W = \oint -p\, dV \qquad [2.5]$$

The work done on the fluid is represented by the area of the cycle in the pressure–volume diagram (Figure 2.5), counted negatively for a clockwise loop and positively for a counter-clockwise loop.

The elementary mechanical work received by the piston (*useful work*) is expressed in the form:

$$\delta W_m = F\, dx \qquad [2.6]$$

where $F = (p - p_a) S$ is the component along the displacement axis x of the force exerted by the connecting rod on the piston. For the whole cycle, we obtain:

$$W_m = \oint (p - p_a) \, S \, dx = \oint -p \, dV + p_a \oint dV \qquad [2.7]$$

and, since the final volume is equal to the initial volume after a cycle:

$$W_m = \oint -p \, dV = W \qquad [2.8]$$

2.2. Theoretical air cycles

2.2.1. *Hypotheses*

The simplifying assumptions are as follows:

– The evolving fluid in the cylinder is an *ideal gas*: the air or the air–fuel mixture during the compression phase and the combustion gases during the expansion are considered to be ideal gases (ideal gases with $c_p = C^{nt}$), with the same thermodynamic properties as air, and the mass of the gases evolving in the cylinder is assumed to be constant during the evolution in a closed system. This assumption is clearly only suitable if the mass of fuel injected remains negligible compared to that of the air.

– The *pressure drops in the intake and exhaust ports are negligible*: it is then possible to assume that the pressure in the cylinder is constant and equal to the ambient pressure p_a during the intake phase, when operating with a full load, or with a constant downstream pressure p_b from the butterfly valve, in the case of a spark-ignition engine with a partial load. Similarly, the pressure in the cylinder is assumed to equal p_a during the exhaust phase. The *heat exchanges throughout the intake and exhaust circuits are neglected*:

– The openings and closings of the valves are assumed to be instantaneous.

– Compression and expansion are assumed to be *adiabatic and reversible*: the heat exchange between the gases, the cylinder walls and the piston is assumed to be negligible.

– The energy released from the combustion is completely transformed into an external heat input.

The different theoretical air cycles differ from each other – in particular, by the type of evolution assumed during the combustion.

2.2.2. Beau de Rochas cycle (Otto cycle)

Combustion is assumed to be carried out *at a constant volume* when the piston is at the top-dead-center; this assumption is suitable for modeling the behavior of a spark-ignition engine.

2.2.2.1. Full load cycle

Figure 2.8 represents the Beau de Rochas cycle as pressure–volume and temperature–specific entropy diagrams.

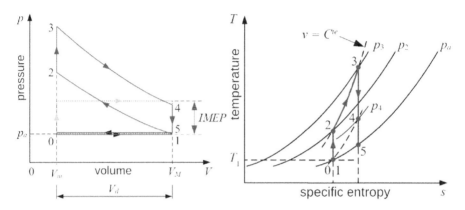

Figure 2.8. *Beau de Rochas' cycle with a full load*

The gases contained in the cylinder at point 0 are the residual gases from the previous cycle. The cycle constitutes the following evolutions:

– Intake (evolution 0–1): the new gases taken in are introduced into the cylinder at atmospheric pressure p_a and temperature T_a. At the end of the intake (point 1), the temperature of the gases is the mixed temperature from the residual gases from the previous cycle and from the new intake gases.

– Compression (evolution 1–2): assumed to be adiabatic and reversible, hence isentropic.

– Combustion (evolution 2–3): the volume remains constant throughout, and the energy is completely transformed into a simple external heat input.

– Expansion (evolution 3–4): assumed to be adiabatic and reversible, hence isentropic.

– Expansion when the exhaust valve opens (evolution 4–5): the opening of the exhaust valve and the resulting expansion to the pressure p_a are assumed to be instantaneous. For the gases remaining in the cylinder, the expansion is assumed to be adiabatic and reversible. During this expansion, the piston displacement is assumed to be zero, and the volume of the space limited by the piston is a constant equal to V_M.

– Exhaust (evolution 5–0): when the expansion phase ends and the exhaust valve opens, the gases are discharged at the constant pressure p_a and temperature T_5.

It should be noted that, unlike the temperature–specific entropy diagram, the pressure–volume diagram is not a state diagram. Indeed, the mass of gas contained in the cylinder is not constant throughout the evolutions 0–1, 4–5 and 5–0. The mass of gas m contained in the cylinder is constant and equal to the mass m_1 after the intake phase in the cycle 1–2–3–4. According to the state equation for ideal gases:

$$m = \frac{p_a V_M}{r T_1} \qquad [2.9]$$

Taking into account $V_d = V_M - V_m$ and the definition of the volumetric ratio τ:

$$V_M = V_d \frac{\tau}{\tau - 1} \qquad V_m = \frac{V_d}{\tau - 1} \qquad [2.10]$$

The work that is involved in the intake and exhaust phases:

$$W_{0 \to 1} = \int_{(0 \to 1)} -p \, dV = -p_a V_d = -W_{5 \to 0} \qquad [2.11]$$

compensate for one other. Hence, we can limit our study of the operating cycle just the evolution phases 1–5.

Since the thermodynamic system considered here is the mass m of gas contained in the cylinder, the compression and expansion transformations are adiabatic, and the heat quantities $Q_{1 \to 2}$ and $Q_{3 \to 4}$ exchanged with the outside are zero. For a closed system and taking into account the ideal gas hypotheses, the work exchanged with the exterior for these evolutions is given by:

$$W_{1 \to 2} = \Delta U_{1 \to 2} = m \, c_v \, (T_2 - T_1) > 0 \qquad [2.12]$$

$$W_{3 \to 4} = \Delta U_{3 \to 4} = m \, c_v \, (T_4 - T_3) < 0 \qquad [2.13]$$

where ΔU corresponds to the variation in the internal energy of the mass of gas m.

The work $W_{2 \to 3}$ and $W_{4 \to 5}$ exchanged during the combustion and exhaust valve opening phases are zero (no piston movement). The combustion energy is assumed to be equivalent to the external heat input $Q_{2 \to 3}$:

$$Q_{2 \to 3} = \Delta U_{2 \to 3} = m \, c_v \, (T_3 - T_2) > 0 \qquad [2.14]$$

The thermodynamic efficiency of the cycle is determined by the mechanical work W_{cycle} exchanged during the cycle:

$$W_{cycle} = W_{1\to 2} + W_{3\to 4} = m\,c_v\,(T_2 - T_1) - m\,c_v\,(T_3 - T_4) \qquad [2.15]$$

along with the amount of heat $Q_{2\to 3}$ generated from the combustion:

$$\eta_{th} = \frac{|W_{cycle}|}{Q_{2\to 3}} = 1 - \frac{T_4 - T_1}{T_3 - T_2} \qquad [2.16]$$

As the compression 1–2 and expansion 3–4 phases are reversible adiabatic transformations involving an ideal gas:

$$\frac{T_2}{T_1} = \frac{T_3}{T_4} = \tau^{\gamma - 1} \qquad [2.17]$$

the thermodynamic efficiency can be written as:

$$\eta_{th} = 1 - \frac{1}{\tau^{\gamma - 1}} \qquad [2.18]$$

Note that the thermodynamic efficiency depends only on the volumetric ratio, and is independent of the quantity of heat provided by combustion, and therefore is also independent of the work produced during the cycle.

The work done by the cycle is given by its area in the pressure–volume diagram. To characterize the engine load in a way that is independent of the displacement, the *indicated mean effective pressure* IMEP is often defined (Figure 2.8), which is the height of the rectangular cycle, equal in area of the actual cycle, such that its base is given by the volume variation V_e:

$$|W_{cycle}| = IMEP \cdot V_d = \eta_{th}\,Q_{2\to 3} \qquad [2.19]$$

Similarly, it is possible to define the *brake mean effective pressure BMEP* by replacing $|W_{cycle}|$ in the definition (equation [2.19]), with the work on the motor shaft per cycle $|W_{shaft}|$.

By introducing the quantity of heat $q_{2\to 3} = Q_{2\to 3}/m$ supplied by combustion per unit mass of air, and by using the Relations [2.9], [2.10], [2.18] and [1.56], the indicated mean pressure can be written as:

$$\frac{IMEP}{p_a} = \frac{q_{2\to 3}}{c_v\,T_1}\,\frac{1 - \dfrac{1}{\tau^{\gamma - 1}}}{(\gamma - 1)\left(1 - \dfrac{1}{\tau}\right)} \qquad [2.20]$$

Gasoline spark-ignition engines operate with sufficiently constant proportions for the air–gasoline mixture, so that all of the oxygen available can be used for the fuel oxidation (see the combustion chapter). Therefore:

$$q_{2\to 3} \approx 2{,}780 \text{ J/kg} \quad \Longrightarrow \quad \frac{q_{2\to 3}}{c_v\, T_1} \approx 13 \quad (T_1 = 25°C) \qquad [2.21]$$

In Figure 2.9, the thermodynamic efficiency of the Beau de Rochas cycle at a full load and the ratio $IMEP/p_a$ are plotted as a function of the volumetric ratio, calculated using the values in equation [2.21]. The thermodynamic efficiency and the indicated mean pressure are increasing functions of the volumetric ratio, and the indicated mean effective pressure increases with the ambient pressure p_a.

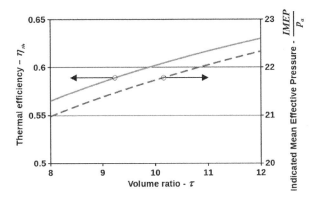

Figure 2.9. *Thermodynamic efficiency and indicated mean pressure of the Beau de Rochas cycle at full load*

2.2.2.2. Partial load cycle

Partially closing the butterfly valve in the intake port causes a pressure drop, and the pressure p_b in the port downstream from the valve is less than the ambient pressure p_a (Figure 2.10).

According to the first law of thermodynamics for a flowing fluid (equation [1.5]), the expansion in the butterfly valve (considered adiabatic) is isenthalpic, and therefore isothermal in the case of a perfect gas (with variations in the kinetic energy ignored). If we assume the pressure in the cylinder remains constant and equal to p_b during the intake phases, then according to the state equation for ideal gases, the mass m_b of gas contained in the cylinder when the intake phases ends (point 1), which remains constant along evolutions 1–2–3–4, is:

$$m_b = \frac{p_b\, V_M}{r\, T_1} \qquad [2.22]$$

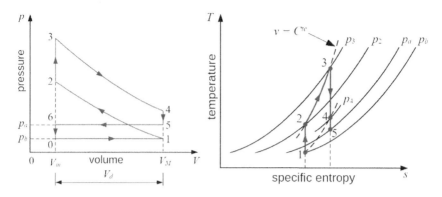

Figure 2.10. *Beau de Rochas cycle at a partial load*

and, assuming that the temperature T_1 remains unchanged as a function of the load:

$$\frac{m_b}{m} = \frac{p_b}{p_a} \qquad [2.23]$$

where m is the mass of gas contained in the cylinder at the end of the intake phase of the cycle at full load.

The evolutions 1–2 and 3–4 are assumed to be adiabatic and reversible, just like the case of a fully loaded cycle, with the mechanical work transferred to the gas in the cylinder being:

$$\begin{aligned}W_{cycle} &= \oint -p\, dV = W_{0\to 1} + W_{1\to 2} + W_{3\to 4} + W_{5\to 6} \\ &= m_b\, c_v\, (T_2 - T_1) + m_b\, c_v\, (T_4 - T_3) + (p_a - p_b) V_d \end{aligned} \qquad [2.24]$$

and, using the reversible adiabatic equation [2.17]:

$$\begin{aligned}|W_{cycle}| &= m_b\, c_v\, (T_3 - T_2)\left(1 - \frac{1}{\tau^{\gamma-1}}\right) - (p_a - p_b) V_d \\ &= m_b\, q_{2\to 3}\left(1 - \frac{1}{\tau^{\gamma-1}}\right) - (p_a - p_b) V_d \end{aligned} \qquad [2.25]$$

In setting $m_b = m$ and $p_b = p_a$, we clearly see that equation [2.25] reduces to that for a fully loaded cycle. In assuming we have the same heat input per unit mass of gas $q_{2\to 3}$, then, from relation [2.23]:

$$\frac{|W_{cycle}|}{|W_{cycle}|_{max}} = \frac{p_b}{p_a} - \left(1 - \frac{p_b}{p_a}\right) \frac{(\gamma - 1)\left(1 - \frac{1}{\tau}\right)}{\frac{q_{2\to 3}}{c_v\, T_1}\left(1 - \frac{1}{\tau^{\gamma-1}}\right)} \qquad [2.26]$$

where $|W_{cycle}|_{max}$ is the work done by the cycle with a full load.

In order for the pressure p_4 at the exhaust valve opening to remain higher than the ambient pressure p_a, the following condition must be satisfied:

$$\frac{p_b}{p_a} > \frac{1}{1 + \frac{1}{\tau^{\gamma-1}} \frac{q_{2\to 3}}{c_v T_1}} \qquad [2.27]$$

The thermodynamic efficiency of the partially loaded cycle is expressed as follows:

$$\eta_{th} = \frac{|W_{cycle}|}{Q_{2\to 3}} = \left(1 - \frac{1}{\tau^{\gamma-1}}\right) - \frac{(p_a - p_b) V_d}{m_b\, q_{2\to 3}} \qquad [2.28]$$

$$\eta_{th} = \left(1 - \frac{1}{\tau^{\gamma-1}}\right) - \left(\frac{1}{\frac{p_b}{p_a}} - 1\right) \frac{(\gamma - 1)\left(1 - \frac{1}{\tau}\right)}{\frac{q_{2\to 3}}{c_v T_1}} \qquad [2.29]$$

Figure 2.11 represents the work done by the Beau de Rochas cycle and its thermodynamic efficiency with a partial load, expressed as functions of the pressure ratio p_b/p_a. A significant degradation of the thermodynamic efficiency is observed for lower loads.

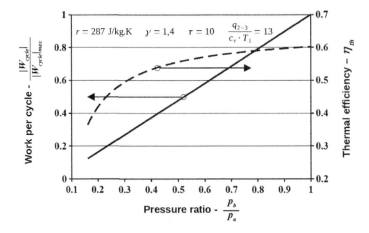

Figure 2.11. *Work done per cycle and thermodynamic efficiency of the Beau de Rochas cycle at a partial load*

2.2.3. Miller–Atkinson cycle

In order to increase the thermodynamic efficiency, there should be more of an expansion than there is compression, and we can shorten the compression phase by delaying the closing time of the intake valve (Figure 2.12). Compression begins when the volume of the cavity delimited by the cylinder and the piston is equal to V'. The pressure and the temperature during the displacement of the piston during the evolution 1–1' are assumed to be constant, and equal to the atmospheric pressure p_a and temperature T_a. The volumetric ratios of the reversible adiabatic compression 1'–2 and expansion 3–4 are, respectively:

$$\tau' = \frac{V'}{V_m} \qquad \tau = \frac{V_M}{V_m} \qquad [2.30]$$

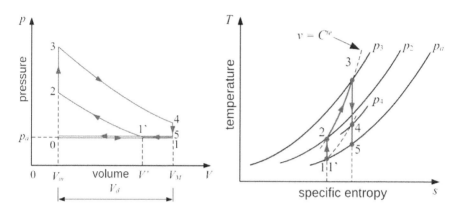

Figure 2.12. *Miller–Atkinson cycle*

The mass of gas m_c contained in the cylinder remains constant throughout the evolution 1'–2–3–4. Assuming $T_1 = T_{1'}$:

$$m_c = m_{1'} = \frac{p_a V'}{r T_1} = \frac{p_a V_d}{r T_1} \frac{\tau'}{\tau - 1} \qquad [2.31]$$

The work done per cycle is given by:

$$|W_{cycle}| = m_c c_v (T_3 - T_4) - m_c c_v (T_2 - T_1) - p_a (V_M - V') \qquad [2.32]$$

and the amount of heat generated by the combustion is:

$$Q_{2\to3} = m_c q_{2\to3} = m_c c_v (T_3 - T_2) \qquad [2.33]$$

Using the reversible adiabatic equation for an ideal gas:

$$T_2 = T_1 \tau'^{\gamma-1} \qquad T_3 = T_4 \tau^{\gamma-1}$$

$$T_3 = T_2 + \frac{q_{2\to3}}{c_v} = T_1 \tau'^{\gamma-1} + \frac{q_{2\to3}}{c_v} \qquad [2.34]$$

we obtain the expression for the thermodynamic efficiency of the cycle:

$$\eta_{th} = 1 - \frac{1}{\tau^{\gamma-1}} - \frac{(\gamma-1)\left(\frac{\tau}{\tau'}-1\right) - \left(1 - \left(\frac{\tau'}{\tau}\right)^{\gamma-1}\right)}{\dfrac{q_{2\to 3}}{c_v\, T_1}} \quad [2.35]$$

in addition to the condition for the pressure p_4 after the expansion phase to be greater than the ambient pressure p_a is:

$$\tau^{\gamma} < \tau'\left(\tau'^{\,\gamma-1} + \frac{q_{2\to 3}}{c_v\, T_1}\right) \quad [2.36]$$

The indicated mean pressure, defined by the expression:

$$IMEP \cdot V_e = \eta_{th}\, m_c\, q_{2\to 3} \quad [2.37]$$

can now be written in the form:

$$\frac{IMEP}{p_a} = \eta_{th}\, \frac{q_{2\to 3}}{c_v\, T_1}\, \frac{\tau'}{(\gamma-1)(\tau-1)} \quad [1.37]$$

Figure 2.13 represents the thermodynamic efficiency and the indicated mean pressure of the Miller–Atkinson cycle as a function of the volumetric ratio τ, given a fixed volumetric ratio τ' of the compression phase. The plot shows that extending the expansion phase increases the thermodynamic efficiency, at the expense of a smaller indicated mean effective pressure.

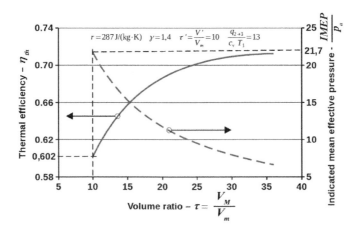

Figure 2.13. *Thermodynamic efficiency and indicated mean pressure as a function of the volumetric ratio τ*

2.2.4. Diesel cycle

The Diesel cycle (Figure 2.14) is similar to the Beau de Rochas cycle, with the exception that the combustion phase 2–3 is assumed to be carried out *at a constant pressure*. The variation in volume during the combustion phase is characterized by the ratio:

$$\mu = \frac{V_3}{V_2} = \frac{V'}{V_m} \qquad V' \leq V_M \Rightarrow \mu \leq \tau \qquad [2.38]$$

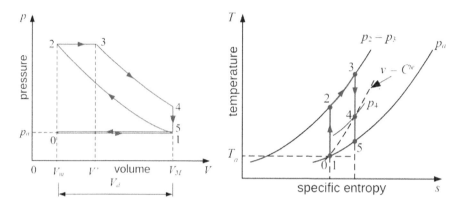

Figure 2.14. *Diesel cycle*

The mass of gas m contained in the cylinder is given by relation [2.9], and is constant during the evolutions 1–2–3–4. The mechanical work exchanged during the cycle is:

$$W_{cycle} = W_{1 \to 2} + W_{2 \to 3} + W_{3 \to 4} \qquad [2.39]$$

where the work done $W_{1 \to 2}$ and $W_{3 \to 4}$ during the compression and expansion phases can be expressed using equations [2.12] and [2.13].

We can use the ideal gas state equation to express the work $W_{2 \to 3}$ exchanged during combustion phase 2–3 as:

$$W_{2 \to 3} = \int_{(2 \to 3)} -p \, dV = -(p_3 V_3 - p_2 V_2) = -m \, r \, (T_3 - T_2) \qquad [2.40]$$

and:

$$\begin{aligned} W_{cycle} &= m \, c_v \, (T_2 - T_1) - m \, r \, (T_3 - T_2) - m \, c_v \, (T_3 - T_4) \\ &= -m \, c_p \, (T_3 - T_2) + m \, c_v \, (T_4 - T_1) \end{aligned} \qquad [2.41]$$

The first law of thermodynamics for a closed system allows us to express the quantity of heat which is produced from the combustion phase as:

$$Q_{2\to 3} = \Delta U_{2\to 3} - W_{2\to 3}$$
$$= m\, c_v\, (T_3 - T_2) + m\, r\, (T_3 - T_2) = m\, c_p\, (T_3 - T_2) \quad [2.42]$$

Now, substituting these two expressions into the definition of the thermodynamic efficiency of the cycle, we obtain:

$$\eta_{th} = \frac{|W_{cycle}|}{Q_{2\to 3}} = 1 - \frac{1}{\gamma}\frac{T_4 - T_1}{T_3 - T_2} \quad [2.43]$$

In using the relationships between the temperatures, the reversible adiabatic equations for compression and expansion, and since combustion occurs at constant pressure:

$$\frac{p_2 V_m}{T_2} = \frac{p_3 V'}{T_3} \text{ with } p_2 = p_3 \Rightarrow T_3 = \mu\, T_2 \quad [2.44]$$

let us write the thermodynamic efficiency of the cycle in the form:

$$\eta_{th} = 1 - \frac{1}{\tau^{\gamma-1}}\frac{1}{\gamma}\frac{\mu^\gamma - 1}{\mu - 1} \quad [2.45]$$

Recalling the indicated mean pressure, defined in equation [2.19]:

$$IMEP \cdot V_e = \eta_{th}\, m\, c_p\, (T_3 - T_2) = \eta_{th}\, m\, \frac{\gamma\, r}{\gamma - 1}(T_3 - T_2) \quad [2.46]$$

can then be given by:

$$\frac{IMEP}{p_a} = \eta_{th}\,(\mu - 1)\, \frac{\gamma}{\gamma - 1}\,\frac{\tau^\gamma}{\tau - 1} \quad [2.47]$$

Figures 2.15 and 2.16 represent the thermodynamic efficiency and the indicated mean pressure of the Diesel cycle, as a function of the volumetric ratio $\tau = V_M/V_m$ and of the volumetric ratio $\mu = V'/V_m$ of the combustion phase.

If $\mu \to 1$, then $V' \to V_m$ and point 3 tends towards point 2. In the case when $\mu = 1$, the compression and expansion phases merge and the work generated by the cycle is zero. When there is no load ($\mu = 1$), the efficiency of the Diesel cycle tends towards that of the Beau de Rochas cycle. When the work generated by the cycle increases (by increasing values of μ), the thermodynamic efficiency of the Diesel cycle decreases. This yield increases with the volumetric ratio τ.

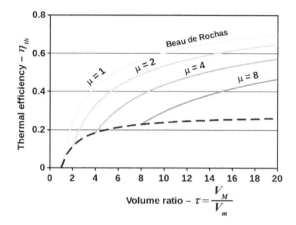

Figure 2.15. *Thermodynamic efficiency as a function of τ and μ*

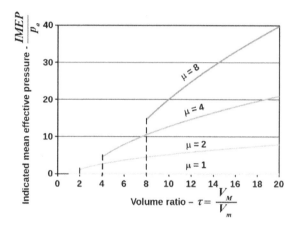

Figure 2.16. *Indicated mean pressure as a function of τ and μ*

2.2.5. *The limited pressure cycle (mixed cycle)*

The constant pressure during the combustion phase in the Diesel cycle does not adequately represent modern compression ignition engines. Furthermore, as we have seen when comparing the Beau de Rochas and Diesel cycles, having a constant volume during combustion is desirable for a higher thermodynamic efficiency, but this would create excessively high pressures for compression ignition engines with a high volumetric ratio. The limited pressure cycle ("Sabathé cycle" in French) is a more accurate model for the operation of compression ignition engines, which breaks down the combustion phase (Figure 2.17) into one with a constant volume (2–3'), and one with a constant pressure (3'–3). This decomposition is determined by the ratios:

$$\nu = \frac{p_3}{p_2} \quad \mu = \frac{V'}{V_m} \quad [2.48]$$

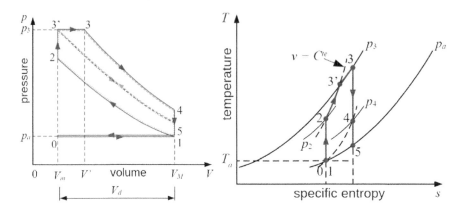

Figure 2.17. *Limited pressure cycle (mixed cycle)*

The limited pressure cycle can be considered as a compromise between the Beau de Rochas and the Diesel cycles, which are plotted in Figure 2.17. The work exchanged during the cycle is:

$$W_{cycle} = W_{1 \to 2} + W_{3' \to 3} + W_{3 \to 4}$$
$$= m\, c_v\, (T_2 - T_1) - m\, r\, (T_3 - T_{3'}) - m\, c_v\, (T_3 - T_4) \quad [2.49]$$

and the amount of heat generated by the combustion:

$$Q_{2 \to 3'} = Q_{2 \to 3} + Q_{3 \to 3'} = m\, c_v\, (T_{3'} - T_2) + m\, c_p\, (T_3 - T_{3'}) \quad [2.50]$$

where m is the (constant) mass of gas, evolving along the phases 1–2–3'–3–4.

By proceeding in a similar way to what was done in the case of the Beau de Rochas and Diesel cycles, we can obtain the following thermodynamic efficiency of the cycle:

$$\eta_{th} = 1 - \frac{1}{\tau^{\gamma-1}} \frac{\mu^\gamma \nu - 1}{\nu - 1 + \gamma\, \nu\, (\mu - 1)} \quad [2.51]$$

With the same volumetric ratio, the performance of the limited pressure cycle is between that of the Beau de Rochas and Diesel cycles.

2.2.6. *Comparison of theoretical air cycles*

– Beau de Rochas, Diesel and the limited pressure cycles (fully loaded):

- The thermodynamic efficiency of the Beau de Rochas cycle depends only on the volumetric ratio τ; it is therefore independent of the work generated by the cycle. The thermodynamic efficiencies of the Diesel and limited pressure cycles decrease when the work supplied from the cycle increases.

- The thermodynamic efficiencies of the three cycles increase with the volumetric ratio.

- For the same volumetric ratio τ and the same quantity of heat released from combustion, the thermodynamic efficiencies are classified as follows: $\eta_{th-\text{Beau de Rochas}} > \eta_{th-\text{limited pressure}} > \eta_{th-\text{Diesel}}$. The thermodynamic efficiency of the Diesel cycle tends towards that of the Beau de Rochas cycle, when the work provided by the cycle tends towards 0. At a normal load and for the same volumetric ratio, the Diesel cycle has an efficiency approximately 15% less than that of the Beau de Rochas cycle. For a given maximum pressure, the optimal efficiency of the limited pressure cycle is obtained for an approximate value of $\mu \approx 2$. This comparison shows that the combustion with a constant volume is preferable in order to optimize the thermodynamic efficiency.

– Beau de Rochas cycle with a partial load: closing the butterfly valve during the intake phase causes a drop in pressure, decreasing the thermodynamic efficiency.

– Miller–Atkinson cycle: a longer expansion phase allows for a greater transformation into mechanical energy, resulting in an improved thermodynamic efficiency, albeit at the cost of a smaller indicated mean effective pressure being generated. For low loads, the operating cycle of some modern engines is close to that of the Miller cycle, and is achieved by offsetting the time when the intake valve opens.

The trends suggested by the theoretical air cycles are consistent with what we have observed for actual internal combustion engines. On the contrary, the estimated temperatures, as well as the calculated efficiencies, are too high.

2.3. Influences of the thermophysical properties of the working fluid on the theoretical cycles

2.3.1. *Thermophysical properties of the working fluid*

For the theoretical cycles that we have been studying, we assumed that the working fluid was dry air modeled as ideal gas in the strict sense: the specific ideal gas constant is the one for air ($r = 287$ J/ (kg · K)), and the heat capacities are assumed to be constant ($c_p = 1,004$ J/(kg · K), $c_v = 717$ J/(kg · K)).

During the operating cycle of an internal combustion engine:

– the temperature fluctuations of the working fluid can exceed 2,000 K, which are highly significant;

– the introduction of the fuel and the chemical reactions of the combustion can induce compositional changes in the working fluid.

For a stoichiometric mixture of air–octane, the theoretical combustion reaction (see section 4.2.2) is written as follows:

$$C_8H_{18} + 12.5\,(O_2 + 3.76\,N_2) \longrightarrow 8\,CO_2 + 9\,H_2O + 47\,N_2 \qquad [2.52]$$

Table 2.2 compares the specific compositions of dry air, the stoichiometric air–octane vapor mixture and the products of their theoretical combustion.

	% O_2	% N_2	% CO_2	% H_2O	% C_8H_{18}
Dry air $r = 287$ J/(kg · K)	23.30	76.70	0	0	0
Mixed air–octane $r = 275$ J/(kg · K)	21.85	71.91	0	0	6.24
Combustion products $r = 286$ J/(kg · K)	0	71.91	19.23	7.86	0

Table 2.2. *Specific compositions of dry air, the stoichiometric air–octane mixture and their theoretical products of combustion*

As the air, air–fuel mixture and the combustion products are assumed to be ideal or semi-ideal gas mixtures, and their specific constants presented in Table 2.2 are given by the equation:

$$r = \frac{\sum_i m_i\, r_i}{\sum_i m_i} \qquad [2.53]$$

where m_i and r_i are the masses and specific constants of the gases that constitute the mixture, respectively.

Similarly at a constant pressure, the specific heat capacity of a gas mixture is given by:

$$c_p(T) = \frac{\sum_i m_i\, c_{pi}(T)}{\sum_i m_i} \qquad [2.54]$$

Figure 2.18 shows the specific heat capacities of dry air, stoichiometric air–octane mixture, and their theoretical combustion products.

The composition of the mixture still weakly influences r, whereas the composition and presence of the constituent fuel (here C_8H_{18}) influences the heat capacities in a non-negligible way. In any case, increasing the temperature gives rise to an increase in the heat capacities.

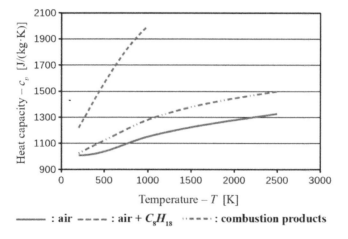

Figure 2.18. *Specific heat capacities (at a constant pressure) of dry air, stoichiometric air–octane mixture and its theoretical combustion products*

2.3.2. *Reversible adiabatic transformations*

The objective here is to analyze the *effects governed by the fluid properties* in the case of a reversible adiabatic transformation, according to the model that characterizes the fluid behavior: as an ideal gas, a semi-ideal gas or a real gas (van der Waals gas).

According to the state equation for a real gas (equation [1.42]):

$$p = \frac{rT}{v-b} - \frac{a}{v^2} \qquad [2.55]$$

where v is the specific volume, p is the pressure, T is the thermodynamic temperature, and r is the specific gas constant. For air, the values used in the real gas equation are the constants $a = 162.1$ Pa·m^6/kg^2 and $b = 1.250 \cdot 10^{-3}$ m^3/kg. It has also been shown that the specific heat capacity at a constant volume of a vans der Waals gas depends solely on its temperature T:

$$c_v = c_v(T) \qquad [1.67]$$

For an isentropic transformation of a real gas ($ds = 0$), we have, according to equation [1.65]:

$$dv = -\frac{c_v(T)}{r}(v-b)\frac{dT}{T} \qquad [2.56]$$

The real gas equation reduces to that of a semi-ideal gas by substituting $a = 0$ and $b = 0$ into it, and may be reduced even further to that of an ideal gas by imposing the additional constraint $c_v(T) = C^{te}$.

Figure 2.19 represents the pressure and temperature at the end of a reversible and adiabatic compression phase as a function of the volumetric ratio τ for air, initially at a pressure of 1.013 bar and a temperature of 25°C, which was calculated using numerical integration techniques applied to equation [2.56] associated with equation [2.55].

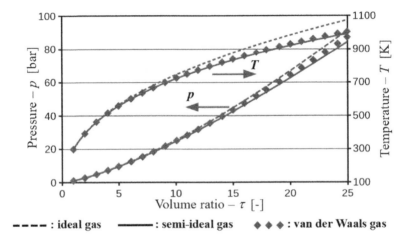

Figure 2.19. *Reversible adiabatic compression: pressure and temperature as a function of the volumetric ratio*

For real and semi-ideal gases, the heat capacity for a constant volume was calculated using equations [1.60] and [1.61], whereas the one used for a ideal gas is that calculated under the initial conditions.

If we apply the state equations to air under different hypotheses, the differences between the final temperatures and pressures remain very low, up to volumetric ratios of the order of 10. Above this, the differences between real and semi-ideal gases then increase, and may reach 82 K and 7 bar for the final temperature and pressure when the volumetric ration reaches 25, respectively. The differences between semi-ideal and real gases remain low in the range of volumetric rations considered here, however.

2.3.3. *Mixed cycle for ideal and semi-ideal gases*

This chapter will analyze how the theoretical cycle is influenced by the nature of the working gas, which we assume to be an *ideal gas* or a *semi-ideal gas*. To be more concrete, we will study the mixed cycle to represent how a naturally aspirated Diesel engine (non-supercharged) operates, whose parameters are the following:
- volume generated by the piston displacement: $V_d = 368$ cm^3;
- volume ratio: $\tau = 22$;
- ambient pressure: $p_a = 1.013$ bar;
- ambient temperature: $T_a = 25°C$ (298 K);
- maximum pressure of the cycle: $p_{max} = 125$ bar;
- combustion heat of intake gas per unit mass: $q_{2\to3} = 1,785$ kJ/kg.

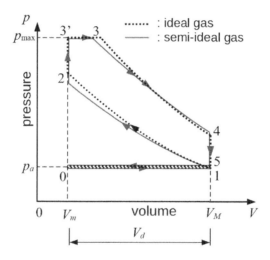

Figure 2.20. *Mixed cycle: comparison of ideal and semi-ideal gas calculations*

Figure 2.20 represents the cycles qualitatively, calculated in assuming that air is an ideal or a semi-ideal gas. With the exception of the hypothesis on the nature of the gas, all the other hypotheses are the same ones used when studying the theoretical air cycles (see section 2.2.1). The specific heat capacities of air are estimated using equations [1.60] and [1.61] (see section 1.3.3 in Chapter 1).

2.3.3.1. *Intake phase (evolution 0–1)*

State 0 (Figure 2.21) corresponds to the end of the exhaust phase (evolution 5–0). Here, the clearance volume V_m contains the residual burnt gases from the previous cycle at ambient pressure $p_a = p_0$ and temperature T_0 (mass $m_0 = (p_0 V_m)/(r T_0)$).

As the pressure drops and heat transfers within the intake port are assumed to be negligible, the volume of gas taken in at ambient pressure p_a and temperature T_a is equal to the volume V_d displaced by the piston (mass $m_a = (p_a\, V_d)/(r\, T_a)$)). State 1 corresponds to the end of the intake phase.

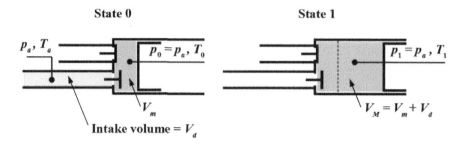

Figure 2.21. *Intake phase*

If we consider the system as consisting of the mass m_0 of the residual gases and the mass m_a of intake gas, and in assuming that the work exchanged by the system with the exterior is zero (constant volume), then the first law of thermodynamics is expressed as the conservation of the internal energy of the system:

$$m_0\, u_0 + m_a\, u_a = (m_0 + m_a)\, u_1$$

$$\Rightarrow m_0 \int_{T_0}^{T_1} c_v(T)\, dT = -m_a \int_{T_a}^{T_1} c_v(T)\, dT \qquad [2.57]$$

Equation [2.57], which allows us to calculate the temperature T_1 for an ideal gas, takes the form:

$$T_1 = \frac{m_0\, T_0 + m_a\, T_a}{m} \quad \text{with: } m = m_0 + m_a \qquad [2.58]$$

2.3.3.2. *Compression phase (evolution 1–2)*

During the compression phase 1–2 with the volumetric ratio τ, both valves are closed. Since we assume that the evolution of a mass m in a closed system is a reversible and adiabatic evolution, from equation [1.57], we have:

$$\int_{T_1}^{T_2} \frac{c_v(T)}{r} \frac{dT}{T} = \ln(\tau) \qquad [2.59]$$

from which, provided we know T_1 and τ, it allows us to determine T_2 and the pressure p_2 from the state equation for a gas ($p_2 = p_1\, \tau\, (T_2/T_1)$).

According to the first law of thermodynamics for a closed system, the work done by the compression is given by:

$$W_{1\to 2} = m(u_2 - u_1) = m \int_{T_1}^{T_2} c_v(T)\, dT \qquad [2.60]$$

2.3.3.3. Combustion phase at a constant volume (evolution 2–3')

Since the maximum pressure p_{max} of the cycle is fixed, the temperature $T_{3'}$ at the end of the combustion phase for a constant volume is obtained directly from the state equation of the working fluid ($T_{3'} = T_2(p_{max}/p_2)$). Applying the first law of thermodynamics with the two states 2 and 3', let us calculate the heat quantity supplied to the fluid by the combustion at constant volume:

$$Q_{2\to 3'} = m(u_{3'} - u_2) = m \int_{T_2}^{T_{3'}} c_v(T)\, dT \qquad [2.61]$$

2.3.3.4. Combustion phase at a constant pressure (evolution 3'–3)

The amount of heat supplied to the working fluid by the combustion at a constant pressure is:

$$Q_{3'\to 3} = m_a\, q_{2\to 3} - Q_{2\to 3'} \qquad [2.62]$$

where $q_{2\to 3}$ is the quantity of heat supplied by the combustion per unit mass of gas taken in, and m_a is the mass of fresh gas taken in per cycle.

The work exchanged with the exterior by the working fluid during the combustion at a constant pressure is given by:

$$W_{3'\to 3} = m \int_{(3'\to 3)} -p\, dV = -p_{max}(V_3 - V_{3'})$$
$$= -m\, r(T_3 - T_{3'}) \qquad [2.63]$$

and, according to the first law of thermodynamics, the quantity of heat supplied by the combustion at constant pressure is:

$$Q_{3'\to 3} = -W_{3'\to 3} + m(u_3 - u_{3'})$$
$$= m\, r(T_3 - T_{3'}) + m \int_{T_{3'}}^{T_3} c_v(T)\, dT = m \int_{T_{3'}}^{T_3} c_p(T)\, dT \qquad [2.64]$$

As $Q_{3'\to 3}$ is known, equation [2.64] allows us to calculate the temperature T_3. Then, the work $W_{3'\to 3}$ can be calculated using equation [2.63] with the volume V_3, using the state equation for the fluid ($V_3 = V_m(T_3/T_{3'})$).

2.3.3.5. *Expansion phase (evolution 3–4)*

We assume that the expansion 3–4 is adiabatic and reversible. The calculation is similar to that done for the compression 1–2. Since the volumetric expansion ratio $\tau' = V_4/V_3$ and temperature T_3 are known, the temperature T_4 at the end of the expansion can be obtained via the equation:

$$\int_{T_4}^{T_3} \frac{c_v(T)}{r} \frac{dT}{T} = \ln(\tau') \qquad [2.65]$$

and the pressure p_4 is then deduced from the state equation for the fluid:

$$p_4 = p_3 \frac{1}{\tau'} \left(\frac{T_4}{T_3}\right) \qquad [2.66]$$

The work generated during the expansion is:

$$W_{3 \to 4} = m(u_4 - u_3) = m \int_{T_3}^{T_4} c_v(T)\, dT \qquad [2.67]$$

2.3.3.6. *Expansion phase when the exhaust valve opens (evolution 4–5)*

During the assumed instantaneous opening of the exhaust valve (point 4), due to the pressure difference between the inside of the cylinder and the exhaust port, some of the gas contained in the cylinder is transferred to the exhaust port with pressure p_a. It is generally accepted that, for the gases remaining in the cylinder, the expansion 4–5 is adiabatic and reversible. It should be noted that the reversibility hypothesis is not suitable for the proportion of the gases that passes into the exhaust port.

The temperature T_5 of the gases in the cylinder at the end of the expansion is obtained by using equation [1.58]:

$$\int_{T_4}^{T_5} \frac{c_p(T)}{r} \frac{dT}{T} = \ln\left(\frac{p_5}{p_4}\right) \qquad [2.68]$$

Since the pressure drops within the exhaust port are neglected throughout the exhaust phase 5–0, we assume that this occurs at a constant pressure and temperature, with the temperature T_0 of the residual gases equal to T_5. The value T_0 may be defined via successive approximations, while still respecting the equality $T_5 = T_0$.

Table 2.3 compares the cycle characteristics that were calculated using the ideal and semi-ideal gas hypotheses, respectively, assuming the conditions discussed at the beginning of section 2.2.3.

In particular, the values in Table 2.3 show that the temperature calculated at the end of combustion (point 3) is lower than that in the case of a semi-ideal gas. The work of the cycle $|W_{cycle}|$ is reduced in the case of the semi-ideal gas, which gives rise to a notable reduction in the thermodynamic efficiency η_{th}.

Point no.	Characteristic [units]	Ideal gas	Semi-ideal gas
0	T_0 [K]	756	819
1	T_1 [K]	306	307
2	T_2 [K] p_2 [bar] $W_{1\to 2}$ [J]	1 050 76.36 238	976 70.98 232
3'	$T_{3'}$ [K] $Q_{2\to 3'}$ [J]	1 719 214	1 721 301
3	T_3 [K] V_3 [cm³] $Q_{3'\to 3}$ [J] $W_{3'\to 3}$ [J]	2 980 30.38 564 -161	2 558 26.04 477 -106
4	T_4 [K] p_4 [bar] $W_{3\to 4}$ [J]	1 083 3.58 -607	1 140 3.76 -608
5	T_5 [K]	756	819
	W_{cycle} [J] η_{th} [-]	-530 0.681	-483 0.621

Table 2.3. *Comparison of the cycle characteristics for ideal and semi-ideal gases*

2.4. Zero-dimensional thermodynamic models

2.4.1. *Hypotheses*

So far, the theoretical air cycles considered do not take into account:

– heat transfers between the working fluids evolving in the cylinder and its walls;

– the opening and closing times of the valves, which limit the mass flow rates of the gas transferred between the cylinders and the intake and exhaust ports;

– the length of the combustion phase, which is determined by a heat release law.

By taking these effects into account, zero-dimensional thermodynamic models give rise to a more realistic description of internal combustion engine cycles.

At any point in the engine cycle of a single-zone model, the properties of the working fluid are assumed to be uniform throughout the cylinder. Multi-zone models (most often two zones) are useful when the fluid in the cylinder cannot be considered as a uniform mixture. In this case, the model is broken down into several volumes, in each of which the properties are uniform but each zone has different values. For example, during the combustion phase in a spark-ignition engine, two zones may be considered: the fresh gas zone and the burnt gas zone.

2.4.2. *Single-zone model*

Figure 2.22 represents the gas transfers between the cylinder and the intake and exhaust ports between two neighboring moments in time, t and $t + dt$. During the time interval dt, the piston moves through the distance $V_p \, dt$ (where V_p is the piston movement speed). m is the mass of the working fluid contained in the cylinder at the time t (where p is the pressure, T is the temperature, u is the specific internal energy, etc.). dm_i is the mass of the intake gas between t and $t + dt$ (where V_i is the gas flow velocity in the intake port, p_i, T_i, u_i, etc.), while the mass dm_e occupying the volume $d\Omega_e$ passes from the cylinder to the exhaust port. At a time $t + dt$, the cylinder characteristics are $m + dm$, $p + dp$, $T + dT$, $u + du$, etc., and the velocities of the gases contained *inside of the cylinder* are assumed to be small enough to be neglected.

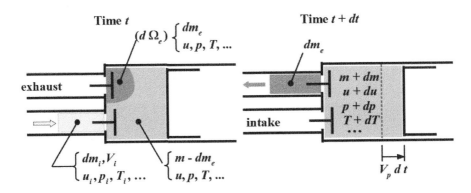

Figure 2.22. *Gas exchange between the cylinder and intake and exhaust ports*

The closed thermodynamic system here consists of the mass $m - dm_e$ of the *gases remaining in the cylinder* between the two moments t and $t + dt$, and also of the *mass dm_i entering into the cylinder*. The conservation of mass for the closed system can then be expressed in the form:

$$dm = dm_i - dm_e = (q_{mi} - q_{me}) \, dt \qquad [2.69]$$

where q_{mi} and q_{me} are the intake and exhaust mass flow rates, respectively.

According to the first law of thermodynamics (equation [1.3]) in its differential form:

$$\delta W_e + \delta Q_e = dU + dE_c \qquad [2.70]$$

In keeping just the first-order terms, the variation in the internal energy of the system can be expressed as follows:

$$dU = U(t+dt) - U(t) = (m+dm)(u+du)$$
$$- [(m-dm_e)u + dm_i\, u_i] = m\,du + dm\,u + dm_e\,u - dm_i\,u_i \qquad [2.71]$$

By neglecting the velocities of the gas in the cylinder, the variation in the kinetic energy is:

$$dE_c = -dm_i \frac{V_i^2}{2} \qquad [2.72]$$

The work δW_e exerted on the system comes from:

– the action of the piston: $-p\,dV$ (where V is the volume of the cylinder/piston cavity);

– the action of the fluid of mass dm_i, upstream in the intake duct: $p_i\,A_i\,V_i\,dt = (p_i/\rho_i)\,dm_i$ (where A_i is the intake duct area, ρ_i is the density);

– the action of the fluid discharged into the exhaust, on the fluid remaining in the cylinder: $-p\,d\Omega_e = -(p/\rho)\,dm_e$.

The heat quantity term δQ_e in equation [2.70] may then be written in the form:

$$\delta Q_e = \Phi_e\,dt \qquad [2.73]$$

where the heat flow Φ_e represents the heat input coming from the combustion, as well as the heat transfers between the wall and the working fluid in contact with it.

By substituting in terms from the expression of the first law (equation [2.70]), and by taking the conservation of mass equation (equation [2.69]) into account:

$$m\frac{du}{dt} = \Phi_e - p\frac{dV}{dt} + (q_{mi} - q_{me})\frac{p}{\rho} - q_{mi}\left(h - h_i - \frac{V_i^2}{2}\right) \qquad [2.74]$$

where h and h_i are the specific enthalpies ($h = u + p/\rho$) of the working fluid in the cylinder and of the intake gases, respectively.

If the working fluid and the intake gases assimilate into a semi-ideal gas then, provided that the engine operates at a constant rotational speed ω as a *stabilized operation*, equations [2.69] and [2.74] then take the form:

$$c_v(T) \frac{dT}{d\alpha} = -\frac{rT}{V}\frac{dV}{d\alpha}$$

$$+ \frac{1}{m\,\omega} \left[\Phi_e + (q_{mi} - q_{me})\, rT - q_{mi}\left(h - h_i - \frac{V_i^2}{2}\right) \right] \quad [2.75]$$

$$\frac{dm}{d\alpha} = \frac{1}{\omega}(q_{mi} - q_{me}) \quad [2.76]$$

where $\alpha = \omega\, t$ is the angle of rotation and $h - h_i = \int_{T_i}^{T} c_p(T)\, dT$. We can find the volume $V(\alpha)$ by studying the kinematics of the piston drive system.

The system of the differential equations [2.75] and [2.76] may be solved numerically, by modeling the flows q_{mi} and q_{me}, as well as the flux Φ_e, which includes a heat production term arising from the combustion, and also a heat exchange term for the walls that are in contact with the working fluid.

2.4.3. *Flow through the valves*

Figure 2.23 represents the geometry of a valve, for example, an intake valve to a cylinder in an engine. A typical expression for a valve opening is also shown, which is a function of crankshaft rotation α. The initial opening and final closing angles of the valve correspond to α_o and α_c, respectively, while the maximum lift is l_{max}.

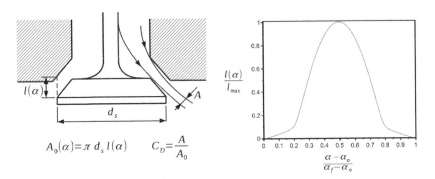

Figure 2.23. *Flow through the valves*

The actual sectional area A of the flow is less than the geometric sectional area $A_0 = \pi \, d_s \, l(\alpha)$, and the flow coefficient can be defined as:

$$C_D = \frac{A(\alpha)}{A_0} \qquad [2.77]$$

In the vicinity of the valve, the flow regime varies according to the lift $l(\alpha)$, with the flow coefficient depending on the latter (Heywood 1988).

Figure 2.24 describes the theoretical model generally used to express the mass flow through a valve. The state i corresponds to the initial state (speed $V_i = 0$) upstream from the valve and is characterized by the pressure p_i, temperature T_i, and density ρ_i. V, p, T, ρ, are the characteristics for the flow in the outlet port and the cross-section of which is A. The reversible adiabatic flow of an ideal gas in a converging nozzle is used as a model.

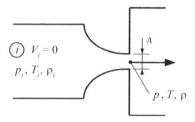

Figure 2.24. *Theoretical model of the flow through a valve*

According to the first law of thermodynamics, the total enthalpy is conserved along a flow:

$$h_i = h + \frac{V^2}{2} \qquad [2.78]$$

which, in the case of an ideal gas, can also be written as:

$$\frac{T_i}{T} = 1 + \frac{\gamma - 1}{2} M^2 \quad \text{with:} \quad M = \frac{V}{c} = \frac{V}{\sqrt{\gamma \, r \, T}} \qquad [2.79]$$

where c is the speed of sound and M is the Mach number.

According to equation [2.79], the flow becomes sonic ($M = 1$) at the nozzle outlet when:

$$\frac{p}{p_i} = \left(\frac{T}{T_i}\right)^{\frac{\gamma}{\gamma-1}} = \left(\frac{2}{\gamma+1}\right)^{\frac{\gamma}{\gamma-1}} \qquad [2.80]$$

In the case of a completely subsonic flow, by relation [2.78] for the velocity calculation, the state equation for the fluid ($p/\rho = r\,T$; $h = c_p\,T$) and the reversible adiabatic law is:

$$q_m = C_D\, A_0\, \frac{p_i}{\sqrt{r\,T_i}}\, \left(\frac{p}{p_i}\right)^{\frac{1}{\gamma}} \sqrt{\frac{2\,\gamma}{\gamma-1}\left[1 - \left(\frac{p}{p_i}\right)^{\frac{\gamma-1}{\gamma}}\right]}$$

[2.81]

$$\text{if } \frac{p}{p_i} > \left(\frac{2}{\gamma+1}\right)^{\frac{\gamma}{\gamma-1}}$$

Provided that the pressure downstream from the nozzle decreases then, when the *sonic conditions are reached*, the mass flow create remains constant and equals that calculated under sonic conditions. Relation [2.81] then simplifies to:

$$q_m = C_D\, A_0\, \frac{p_i}{\sqrt{r\,T_i}}\, \sqrt{\gamma \left(\frac{2}{\gamma+1}\right)^{\frac{\gamma+1}{\gamma-1}}}$$

[2.82]

2.4.4. *Heat transfer with the cylinder walls*

There are different correlations used in the literature to estimate the *heat flux density* φ (heat flux per unit area) that is exchanged between the working fluid and the walls which it is in contact with which takes the form:

$$\varphi = h\,(T - T_w)$$

[2.83]

where T and T_w are the fluid and wall temperatures, respectively, and h is the exchange coefficient.

Woschni's correlation (Heywood 1988) is the one used by many authors. The starting point that Woschni used to express the convection exchange coefficient was the correlation $Nu = 0.035\,Re^m$, where the Nusselt number and the Reynolds number ($Nu = (h\,D)/k$; $Re = (\rho\,w\,D)/\mu$), are defined by taking the diameter D of the cylinder as the reference dimension. Here, w is the average gas velocity in the cylinder. For a gas that obeys the ideal gas state equation $p/\rho = r\,T$, variations in the thermal conductivity k and dynamic viscosity μ (as a function of the temperature T) are estimated to be $k \sim T^{0.75}$ and $\mu \sim T^{0.62}$. During the gas exchange and compression phases, the average gas velocity $\overline{w}$ in the cylinder is assumed to be proportional to the average velocity $\overline{V}_p$ of the piston, and a complementary term is added to the velocity to take into account the density variations induced by combustion. In the absence of "swirl" (fluid rotation within the cylinder), the velocity w can be expressed in the form:

$$w = C_1\,\overline{V}_p + C_2\, \frac{V_d\, T_r}{p_r\, V_r}\,(p - p_m)$$

[2.84]

where V_d is the volume displaced by the piston movement, p_r, V_r, T_r are the pressure, volume and temperature under the chosen reference conditions (e.g. when the intake phase ends), p is the pressure in the cylinder, and p_m is the pressure that would exist in the cylinder in the absence of combustion ($p_m = p_r \left(V_r/V\right)^\gamma$). The constants C_1 and C_2 are listed in Table 2.4.

Phase	C_1	C_2
Intake Exhaust	6.18	0
Compression	2.28	0
Combustion Expansion	2.28	3.24×10^{-3}

Table 2.4. *Coefficients used to calculate the velocity w [m/s]*

The convective exchange coefficient is given by the relationship:

$$h_{[W/(m^2 K)]} = 130 \, D_{[m]}^{-0.2} \, p_{[bar]}^{0.8} \, T_{[K]}^{-0.55} \, w_{[m/s]}^{0.8} \qquad [2.85]$$

The *radiation heat exchanges* remain quite low compared to the convection exchange in spark-ignition engines. On the contrary, in compression ignition engines this mode of transfer is non-negligible. The radiation term proposed by Annand (1963) is:

$$h = a \, \frac{k}{D} \, Re^{0.7} + c \, \sigma \, \frac{T^4 - T_w^4}{T - T_w}. \qquad [2.86]$$

The first term corresponds to convection exchanges, and is expressed as a function of the Reynolds number ($Re = (\rho \, \overline{V}_p \, D)/\mu$), and the second term corresponds to radiation heat exchanges. The constant a here takes values between 0.35 and 0.8. The proposed values for the constant c, which is zero for the phases other than combustion and expansion, are as follows:

– $c = 0.576$ for a compression ignition engine;

– $c = 0.075$ for a spark-ignition engine.

Here, σ is the Stefan–Boltzmann constant ($\sigma = 5.67 \times 10^{-8}$ W/(m$^2 \cdot$ K^4)).

2.4.5. *Combustion heat generation model*

The Wiebe function Ghojel (2010), which expresses the *fraction of burnt gas* $x_b = m_b/m$ as a function of the crankshaft rotation angle α (where m is the total mass of the gas contained in the cylinder, m_b is the mass of the burnt gas), is used to model the heat release law for the combustion:

$$x_b = 1 - \exp\left[-a\left(\frac{\alpha - \alpha_0}{\Delta\alpha}\right)^{n+1}\right] \qquad (\alpha_0 \leq \alpha \leq \alpha_0 + \Delta\alpha) \qquad [2.87]$$

where α_0 is the initial combustion angle and $\Delta\alpha$ is the angular combustion duration.

Assuming that combustion finishes when $x_b = 99.9\%$, we obtain for the value of the constant $a = 6.908$.

The mass of gas burned per unit angle is given by:

$$\frac{dm_b}{d\alpha} = \frac{m}{\Delta\alpha} a(n+1) \left(\frac{\alpha - \alpha_0}{\Delta\alpha}\right)^n \exp\left[-a\left(\frac{\alpha - \alpha_0}{\Delta\alpha}\right)^{n+1}\right] \qquad [2.88]$$

Figure 2.25 shows how the distribution of heat release evolves according to Wiebe's law, as well as its derivative with respect to the crankshaft rotation angle as a function of the rotation angle α, which depends on the choice of parameter n.

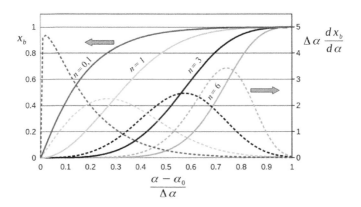

Figure 2.25. *Wiebe's law*

The injection phase in a diesel engine is generally carried out in two stages: pre-injection of a small quantity of diesel to initiate combustion, followed by the main injection. In this case, we can model the heat release such as that represented in Figure 2.26, by using Wiebe's double law with a suitable choice of parameters. The parameters α_{01} and $\Delta\alpha_1$ correspond to the *pre-injection*, while α_{02} and $\Delta\alpha_2$ are those for the *main injection*.

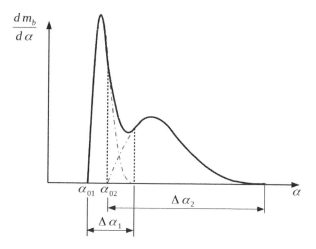

Figure 2.26. *Wiebe's double law*

2.4.6. *Two-zone model*

The multi-zone approach aims to model situations in which assuming the fluid properties are uniform are not realistic. The most commonly used is the two-zone model, and it is particularly useful when representing the evolutions in the cylinder during the combustion phase of a spark-ignition engine (Lounici et al. 2011).

In this case (Figure 2.27), the volume of the cylinder is divided into two zones: the burnt gas zone (volume V_b, mass m_b) and the fresh gas zone (volume V_f, mass m_f). In each of the two zones, the fluid properties are assumed to be uniform throughout, and the zones are assumed to be in a pressure equilibrium with one another. dm_b is the mass of gas from the intake gases that then undergo combustion between the moments t and $t + dt$. During the combustion phase, the valves remain in the closed position, and it is assumed that there is no gas exchange between the cylinder and the intake and exhaust ports. Leaks between the piston and cylinder ("blow-by") will also be neglected. Under these conditions, the total mass $m = m_b + m_f$ of gas contained in the cylinder remains constant.

Assuming that the intake gases and combustion products are ideal or semi-ideal gases:

$$p_f V_f = m_f r_f T_f \qquad p_b V_b = m_b r_b T_b \qquad [2.89]$$

where r_f, T_f, and r_b, T_b are the specific perfect gas constants and temperatures of the intake gases and combustion products.

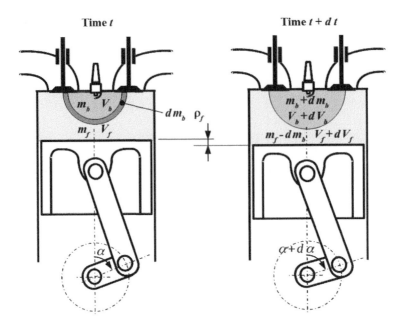

Figure 2.27. *Two-zone model*

The thermodynamic system considered at time t consists of the mass m_b (volume V_b) of burnt gas, and the mass dm_b of the intake gas undergoing combustion between times t and $t + dt$, which occupies the volume dm_b/ρ_f at the time t. At a time $t + dt$, the system under study occupies the volume $V_b + dV_b$. According to the first law of thermodynamics:

$$-p\left[(V_b + dV_b) - \left(V_b + \frac{dm_b}{\rho_f}\right)\right] + \sum \delta Q_{eb} = \qquad [2.90]$$
$$(m_b + dm_b)(u_b + du_b) - (m_b\, u_b + dm_b\, u_f)$$

where $\sum \delta Q_{eb}$ represents the heat exchange between the gases and the walls, as well as the contributions coming from combustion. u_b and u_f are the specific internal energies of the burnt and fresh gases, respectively.

After simplifying and keeping just the first-order terms:

$$\frac{d(m_b\, u_b)}{d\alpha} = -p\frac{dV_b}{d\alpha} + \sum \frac{\delta Q_{eb}}{d\alpha} + \frac{dm_b}{d\alpha} h_f \qquad [2.91]$$

where h_f is the specific enthalpy of the fresh gases.

Similarly, considering the intake gases between the moments t and $t + dt$:

$$\frac{d(m_f u_f)}{d\alpha} = -p\frac{dV_f}{d\alpha} + \sum \frac{\delta Q_{ef}}{d\alpha} - \frac{dm_b}{d\alpha} h_f \quad [2.92]$$

The elementary work exchanged between the gases contained in the cylinder and the piston is then:

$$\delta W = -p\, dV \quad \text{with} \quad dV = dV_b + dV_f \quad [2.93]$$

2.5. Supercharging of internal combustion engines

2.5.1. *Basic principles of supercharging*

The objective of supercharging is to *increase the engine power* by increasing the mass of the working fluid contained in the cylinder, which can be obtained by either by:

– increasing the inlet pressure;

– cooling the air taken in.

Various processes make it possible to supercharge internal combustion engines, such as:

– *pressure wave* effects in the intake port that promote the filling of the cylinder, and burnt gas discharge in the exhaust port;

– *mechanically driven compressors* (Figure 2.28(a)). The compressor can be either a positive displacement pump (roots, rotary-screw, etc.), or a turbomachine powered directly from either the motor shaft or indirectly by an electric motor;

– a *turbocharger* unit (Figure 2.28(b)). The turbine driving the compressor extracts the required energy from the exhaust gases. It should be noted that the presence of a turbine at the exhaust causes back pressure, which has ramifications for the engine cycle.

The air cooler is not present in all superchargers. The air cooling that occurs after the compression causes the density of the intake air to increase, therefore increasing the mass of working fluid contained in the cylinder after the intake phase.

In the case of motor vehicles, supercharging allows for the displacement within the engine to be reduced ("downsizing"), thus making it possible to minimize mechanical losses for low loads and, for spark-ignition engines, the losses produced within the intake port when closing the butterfly valve, overall improving fuel consumption.

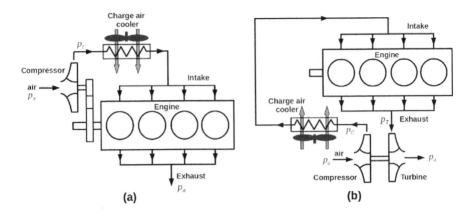

Figure 2.28. *Supercharging in internal combustion engines (a) Mechanically driven compressor and (b) turbocharger*

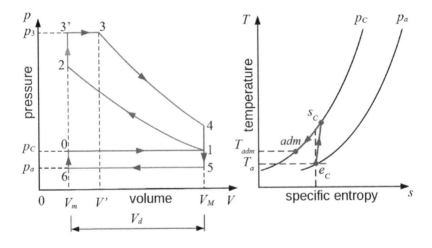

Figure 2.29. *Supercharging by a driven compressor*

2.5.2. Supercharging by a driven compressor

The pressure–volume $p - V$ diagram in Figure 2.29 represents the theoretical changes within the cylinder of an engine supercharged by a driven compressor. Here, p_a and T_a are the pressure and temperature at ambient conditions, respectively, and p_C is the compressor discharge pressure. The pressure drops in the air cooler and the intake port pipes are neglected. Point 6 corresponds to the closing of the exhaust valve and the opening of the intake valve. The cylinder is then connected to the compressor exhaust, and the pressure passes from the pressure p_a to the pressure p_C. It is assumed that the intake phase 0–1 is carried out at the constant pressure p_C. Between the closing of the intake valve (point 1) and the opening of the exhaust valve (point 4), analogously to the case of naturally aspirated engines, the compression (1–2), combustion (2– 3'–3) and expansion (3–4), happen for a constant mass of working fluid. When the exhaust valve opens (point 4), the cylinder is exposed to the atmosphere, and the pressure in the cylinder goes from p_4 at the end of the expansion to the ambient pressure p_a. Discharging the burnt gases occurs at a constant pressure equal to p_a, as we neglect the pressure drops in the exhaust port.

The points e_C and s_C of the temperature–mass entropy $T - s$ diagram in Figure 2.29 represent the air properties during the intake and discharge phases of the compressor. Air is taken in at ambient conditions p_a, T_a. The air cooler lowers the air temperature from the compressor outlet temperature to T_{adm}, thus increasing the air density. The work done on the shaft per unit mass of air w_{aC} that is needed to drive the compressor (considering air as an ideal gas) is:

$$w_{aC} = \frac{w_{iC}}{\eta_{0C}} = \frac{h_{s_C} - h_{e_C}}{\eta_{0C}} = \frac{c_p \left(T_{s_C} - T_{e_C}\right)}{\eta_{0C}} \quad [2.94]$$

where η_{0C} is the mechanical efficiency of the compressor; $h_{e_C}, T_{e_C}, h_{s_C}, T_{s_C}$ are the specific enthalpies and temperatures at the compressor inlet and outlet, respectively, and w_{iC} is the indicated specific work from the compressor.

2.5.3. *Turbocharging*

The pressure–volume diagram in Figure 2.30 represents the theoretical evolution within the cylinder of an internal combustion engine which is supercharged by a turbocharger unit. The pressure within the intake port downstream from the compressor is assumed to be constant, and equal to the exhaust pressure p_C of the compressor, and the pressure in the exhaust port upstream from the turbine is p_T, also assumed to be constant. When the exhaust valve closes and the inlet valve opens (point 6), the pressure in the cylinder goes from p_T to p_C, and then the intake continues at the constant pressure p_T. After closing the inlet valve, the closed system evolution 1–2–3'–3–4 are similar to those already described when we considered the case of a driven compressor supercharger. When the exhaust valve opens (point 4), the fluid pressure decreases from p_4 to p_T, with then the discharge continuing until point 6 at constant pressure p_T.

The evolution that occurs during the intake and exhaust phases is presented in the entropy diagram of Figure 2.30. The evolution of the air within the compressor $(e_C - s_C)$ and in the air cooler $(s_C - adm)$ is similar to that observed for a driven compressor. In the turbine, the expansion $e_T - s_T$ provides the work w_{aT} required to drive the compressor:

$$|w_{aT}| = \eta_{0T} |w_{iT}| = \eta_{0T} (h_{e_T} - h_{s_T}) = \eta_{0T} c_p (T_{e_T} - T_{s_T}) \quad [2.95]$$

where η_{0T} is the mechanical efficiency of the turbine; $h_{e_T}, T_{e_T}, h_{s_T}, T_{s_T}$ are the specific enthalpies and temperatures at the turbine inlet and the outlet, respectively.

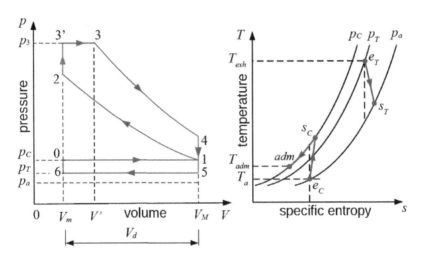

Figure 2.30. *Turbocharging*

Since the role of the turbine is to drive the compressor, the characteristics of the exhaust gases at the inlet of the turbine must be determined, to check whether the energy potential of these gases is sufficient to perform this function. In order to estimate the theoretical temperature T_T of the exhaust gases (assumed to be perfectly mixed at the turbine intake), the considered system starts when the expansion phase (point 4) ends, and finishes at the end of the exhaust phase (point 6). It system consists of the mass m of gas, assimilated into a perfect gas which is contained in the cylinder at point 4 (Figure 2.31).

The gases contained in the cylinder when the expansion ends (point 4) have the following characteristics: a known mass m, pressure p_4 and temperature T_4. The exhaust pipe pressure p_T is assumed to be constant. When the exhaust valve opens, the drop between the pressures p_4 and p_T (evolution 4–5) follows from the fraction of the gases in the cylinder which is discharged into the exhaust pipe. For the gases remaining in the cylinder (mass m_5), we assume a following reversible adiabatic transformation hypothesis:

$$T_5 = T_4 \left(\frac{p_5}{p_4}\right)^{\frac{\gamma-1}{\gamma}} = T_4 \left(\frac{p_T}{p_4}\right)^{\frac{\gamma-1}{\gamma}} \quad [2.96]$$

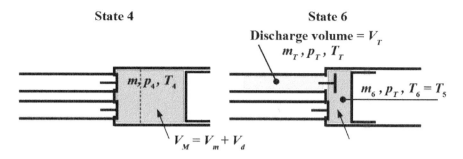

Figure 2.31. *Exhaust gas transfer*

For the gases leaving the cylinder during evolution 4–5, the transformation is assumed to be adiabatic but, after taking into account the throttling losses within the exhaust valve passage, the reversibility assumption is not realistic. The gas volume V_d still present in the cylinder at point 5 is then discharged at a constant pressure and temperature ($T_6 = T_5$) towards the exhaust pipe. At the end of the exhaust phase, the residual mass of gas contained in the cylinder is:

$$m_6 = \frac{p_T V_m}{r T_6} = \frac{p_T V_m}{r T_5} \qquad [2.97]$$

If we neglect the kinetic energy of the gases in the exhaust pipe, then, according to the first law of thermodynamics between States 4 and 6:

$$p_T V_d - p_T V_T = (m_6 u_6 + m_T u_T) - m u_4 \qquad [2.98]$$

where $p_T V_d$ is the work exerted by the piston on the closed system, $-p_T V_T$ is the work exerted by the exhaust pipe gases on the system (V_T is the exhaust volume discharged per cycle), and u_4, u_6, u_T are the specific internal energies at points 4, 6, and finally, in the exhaust pipe (with temperature T_T).

By considering the ideal gas state equation in equation [2.98], we have:

$$T_T = \frac{p_T V_d + m_T c_v T_4 - m_6 c_v T_5}{m_T c_p} \qquad [2.99]$$

Figure 2.32 represents the entropy diagram in which the exergy ex_T of the exhaust gases (equation [1.32]), represented as an area, assumes the conditions during the turbine intake phase. That is, the amount of the exhaust gas energy which can be converted into mechanical energy by reversibly bringing these gases back to the reference state 0.

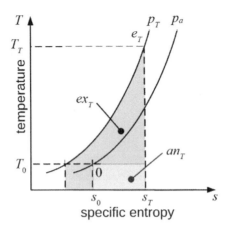

Figure 2.32. *Exhaust gas exergy*

With respect to the reference state, the enthalpy of the exhaust gases $h - h_0$, represented by the entire area located under the isobaric curve p_T between the temperatures T_0 and T_T, cannot therefore be completely converted into mechanical work, and the anergy an_T corresponds to the part of this enthalpy that cannot be converted into mechanical energy.

2.6. Conclusions and perspectives

Theoretical air cycles overestimate the performance and efficiency of internal combustion engines; nevertheless, the *behavior* they predict are satisfactory. The increase in the volumetric compression ratio and constant volume combustion are favorable with regard to heat efficiency, but these parameters are subject to certain constraints: limitations placed on the volumetric ratio to prevent the combustible mixture from self-igniting in the case of spark-ignition engines, the maximum pressure being too high after combusting at a constant volume in the case of Diesel engines, etc. However, improvements are still possible, such as having the operating cycle being close to that of the Miller–Atkinson cycle, which is energetically more efficient with the gas by extending the expansion phase. This can be obtained by shifting the cycle timings, and which can be further combined with supercharging – an improvement that is already implemented in automotive engines.

Among their various evolutions, variable displacement engines are limited by their mechanical complexity as well the opening and closing of the valves ("camless" engines), controlled electromagnetically. This makes it possible to remove the geometric constraints of the camshafts, which are difficult to fine-tune.

The *zero-dimensional models*, with one or more zones, make it possible to represent fairly precisely the operation of internal combustion engines. Their major drawback is that they require values for the coefficients, which are necessary for the heat transfer and combustion heat release models, and require experimental validation.

The quantity of the exhaust gas energy which can be converted into mechanical energy is characterized by its *exergy*. This conversion into mechanical energy is used to power the turbine, driving the compressor in the turbocharger supercharging system, and also in the *waste heat* recovery system using a *combined cycle* (ORC: Organic Rankine cycle). The thermal energy from the exhaust gases is used as a heat source for *cogeneration* systems. We will go into further detail on these points in Chapter 6.

Numerical modeling (CFD) may be implemented to study the local effects during the cycles within an internal combustion engine cylinder, but requires significant energy and computer resources given the complexity of the situation, for example, volume variability of the cylinder–piston cavity, complex flows with heat transfers, intake and exhaust pipe coupling, and chemical combustion reactions.

Research work on high-efficiency gasoline engines and the use of new fuels is being developed with an efficiency target of around 50%. With improving yields, the usage of hydrogen as a fuel in an internal combustion engine (Largeron 2021) is a solution that was originally ruled out. Now however, with more advanced technology that can be implemented quickly, it is finding a renewed interest with the advent of fuel cells, in particular for motorizing heavy vehicles. The impact on CO_2 emissions coming from hydrogen use clearly depends on its low-carbon production.

2.7. References

Annand, W.J.D. (1963). Heat transfer in the cylinders of reciprocating internal combustion engine. *Proceedings of the Institution of Mechanical Engineers*, 177, 973–990.

Fayette Taylor, C. (1985). *The Internal Combustion Engine in Theory and Practice*, 2nd edition. The MIT Press, Cambridge, MA.

Ghojel, J.I. (2010). Review of the development and applications of the Wiebe function: A tribute to the contribution of Ivan Wiebe to engine research. *International Journal of Engine Research*, 11, 297–312.

Heywood, J.B. (1988). *Internal Combustion Engine Fundamentals*. McGraw-Hill, New York.

Largeron, D. (2021). En pointe, l'IFPen Lyon développe l'utilisation de l'hydrogène là où on ne l'attendait pas : dans les moteurs thermiques ! *Lyon Entreprises, Newsletter*, 22 March.

Lounici, M.S., Loubar, K., Balistrou, M., Tazerout, M. (2011). Investigation on heat transfer evaluation for a more efficient two-zone combustion model in the case of natural gas SI engines. *Applied Thermal Engineering*, 31, 319–328.

Pulkrabek, W.W. (2020). *Engineering Fundamentals of the Internal Combustion Engine*, 2nd edition. Prentice Hall, Hoboken.

3

Aeronautical and Space Propulsion

Yannick MULLER and François COTTIER

MTU Aero Engines, Munich, Germany

3.1. History and development of aeronautical means of propulsion

When we think of the origins of human flight, many authors recall the Ancient Greek myth of Daedalus, inventor, sculptor, architect and blacksmith. Along with his son *Icarus*, he made wings out of feathers, wood and wax, to escape the labyrinth of King Minos.

The idea of muscular flight continued throughout the centuries, from antiquity to *Leonardo da Vinci*, most recently with the world record attempt in 1988 by the Daedalus aircraft (so named in honor of the precursor).

It is important, however, to note that the development of aeronautics really started when the primordial idea of the imitation of nature and bird flight was challenged.

Hence, we witnessed the first hot air balloons towards the end of the 18th century (the first hot air balloon flight was completed by the *Montgolfier* brothers, with human passengers on November 21, 1783), whose development then led to airships in the first half of the 20th century.

The first parachute jump from a hot air balloon was in 1797 by the Frenchman *André Jaques Garnerin*.

Next comes (fixed-wing) gliders, with pioneers such as Otto Lilienthal in Germany during the second half of the 19th century.

The first attempts to motorize flight were done by *Clément Ader* and his Éole on October 9, 1890, followed by his Avion III which completed a flight with a horizontal distance of 300 m in 1897. Nevertheless, the first motorized aircraft based on the steam engine had a power-to-weight ratio which was far too low.

Motorized flight experienced a considerable development with the introduction of internal combustion engines (the four-stroke patent by the German *Nikolaus Otto* in 1864). The first propelled and controlled flight in aeronautic history which can be attested was completed by the brothers, *Orville and Wilbur Wright*, with their Flyer on December 17, 1903.

From that moment on, the development of an evermore efficient motorization (power-to-weight ratio) and progress in aeronautical construction (lightness, robustness, aerodynamics) started to bloom.

In under just one year after their first flight, the Wright brothers performed the first closed circle flight on September 20, 1904. In 1906, billionaire *Santos Dumont* performed the first flight in Europe and, on July 25, 1909, *Louis Blériot* crossed the English channel at the controls of his Blériot XI.

May 13, 1913 marks the first flight of a four-engine aircraft, designed by the Russian *Igor Sikorsky*, and on July 23 in the same year, *Roland Garros* crossed the Mediterranean Sea.

Aircraft and engines were gaining robustness and reliability; at the same time, the increase in engine power allowed for higher speeds, which required the use of fairings (dressings of the structure) to increase the aerodynamics. On December 12, 1915, the first fully metal aircraft, the Junkers J 1, took off. The search for increasingly better aerodynamics meant that the radial engine configuration was gradually abandoned, in favor of inline (or straight) engine configurations. These latter configurations possessed a much smaller frontal surface area when compared to their predecessors, although with the trade-off that they can no longer be cooled using the air circulation around the cylinder – water cooling was required instead.

World War I saw the deployment of fighter aircraft and bombers, which naturally required much more engine power for either higher speeds or to carry larger payloads.

The inter-war years saw the advent as well as the decline of diesel engine propelled airships (May 7, 1937: the Hindenburg disaster). Wood and canvas materials were replaced with metal structures and cockpits, which were required for greater powers, continued to improve the engines (radial, inline, V engines). With the development of wireless radio, avionics (altitude indicators in 1914) and passenger comforts for commercial flights, engines also needed to provide the necessary energy for the proper functioning of the aircraft.

World War II saw an even greater engine power developed, to propel aircraft much faster (hunters) or higher, to carry heavier payloads (bombers, transport) and to power much more energy-hungry systems (radio, the first navigation aids, electrical systems, heating, pressurization for high altitude flight). However, it should be noted that the maximum speed of aircraft powered by propeller engines was limited to a Mach number of about 0.7. This limit was imposed by the speed of the blade end, which has to remain in the subsonic region throughout its range of operation.

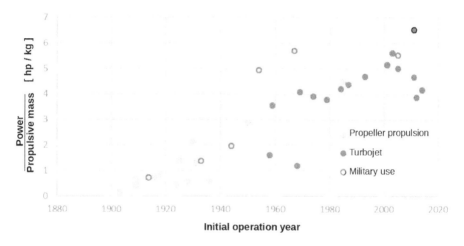

Figure 3.1. *Evolution of the power-to-mass ratio during the 20th century*

Towards the end of the 1930s, development was ongoing in parallel, in both England (*Frank Wittle*) and in Germany (*Papst von Hohain*):

– the first jet aircraft, the Heinkel He 178, was propelled by two rocket engines and performed its maiden flight on August 27, 1937;

– the first jet engines to equip the Gloster E.28/39 were two Rolls-Royce W.2b/Welland MK.I engines (7.55 kN of unit thrust), and the Messerschmitt 262 was powered by two Junkers Jumo 004 B engines (8.8 kN of unit thrust), the first jet airplane to operate in 1944 during World War II.

After World War II, the jet engine would gradually dethrone the piston engine for all military activities (fighting, reconnaissance, bombing) and civil transport. However, it remains largely used in light civil aviation.

The jet engine, or turbojet, underwent several improvements to increase its power (fighter jet aviation), reduce its consumption (civil and military transport), improve its reliability and be available in many variants of increasing complexity.

Research within the aerospace industry was developed in parallel with that of the piston engine, then turbojets. Thus, in France during World War I, there were no less than 19 aircraft manufacturers, though according to bankruptcies and mergers, only two still remain today: Dassault Aviation and Airbus.

The same phenomenon was observed in the aeronautical propulsion industry. Nowadays, four major companies dominate the global market: General Electrics (GE aviation, AVIO) and United Technologies (UTC, Pratt & Wittney) in the United States, Rolls Royce (R&R, ITP) and Safran (Snecma, Turbomeca) in Europe. Many subcontractors such as GKN or MTU Aero Engines, IHI revolve around larger groups, specializing in the development and production of engine subsystems.

At present, by a strategic and economic choice, no civil engine is the work of a single company. In the most widespread model, subcontractors ensure the development of subsystems in which they are specialized; the engine manufacturer ensures the integration of the engine and interface with the manufacturer of aircraft. This model reduces the risks and costs for each partners (Risk and Revenue Sharing Partners).

For reasons of national sovereignty, cooperation is more restricted for international military engines, even if the RRSP model tends to spread (TP400, MTR390, EJ200, T800).

3.2. Presentation of the aircraft system and its propulsive unit

3.2.1. *Classification and presentation of the usual architectures of aeronautical engines and their specific uses*

It is possible to classify aeronautical engines (Bauer 2009; Bouchez 2018) into two large groups (Figure 3.2).

The first of these groups are the *air-breathing engines* or *aerobic* (which use the oxygen in the ambient air as an oxidizer to ensure fuel combustion), and the second are the *anaerobic motors* (which carry their own oxidizer, hence are unaffected by variations in the atmospheric oxygen concentration). The second group brings together a large number of rocket engines covered later in this chapter.

The group of aerobic engines further subdivides into two large groups: the first brings together internal combustion engines, or piston engines, whose operation is identical to the four-stroke engine used in automobile engines, which differs in that the engine drives a propeller with a fixed or variable pitch. This type of engine is usually used in leisure or light aviation.

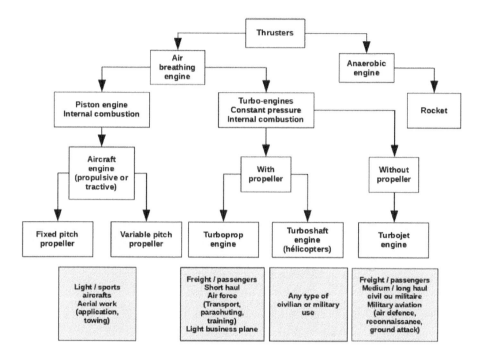

Figure 3.2. *Aeronautic engine configurations and their applications*

Finally, the second large group consists of constant pressure combustion engines, or turbomachines.

When these latter engines drive a propeller external to the engine, we say that they are *turboprop* engines when used in fixed-wing aircraft (aircraft) and *turboshaft* engines when used in rotating-wing aircraft (helicopters).

This type of turbomachine is used in short/medium transport (ATR42) and long-distance flight (A400M, C-130 tactical military transport), as well as in helicopters of all types: civilian and military.

When a turbomachine drives a ducted propeller (inside of the nacelle), we say that it is a *turbojet*. The vast majority of freight and passenger aircraft, as well as modern combat aircraft, use this type of engine.

Whichever configuration is used for a civilian, military or industrial application, a turbojet engine (Farokhi 2014; Sforza 2016) (Figure 3.3) always consists of five main elements:

– an air intake;
– a compressor (axial or radial);
– a combustion chamber;
– a turbine;
– a hot gas ejection nozzle.

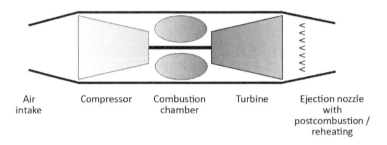

Figure 3.3. *Configuration of a single-spool turbojet*

The purpose of the air inlet is to bring the air to the entrance of the compressor, in the most uniform and homogeneous way possible. The compressor provides the necessary work to compress the air into the combustion chamber, where the energy level of the air is increased by the calorific value of the fuel burned. The combustion gases are then expanded in the turbine, which is connected to the compressor by a shaft. The purpose of the turbine is to convert part of the internal energy of the hot gases into mechanical rotational energy to set the compressor in motion. The remainder of the energy of the gas will be converted into kinetic energy within the nozzle to ensure surplus air velocity at the outlet of the nozzle relative to the air entering the turbomachine in the case of a turbojet engine.

In the case of a turboprop, it is necessary to extract more energy from the gases in the turbine, in order to additionally rotate the propeller.

The upstream portion of the combustion chamber (air intake, compressor) is called the *cold section*, whereas the downstream part which makes up the engine (turbine, nozzle) is called the *hot section*.

The first aeronautical turbomachines in their original form were *simple-cycle single-spool* turbojets, for which the entire airflow that enters into the engine passes through the compressor connected to the turbine by a single shaft.

Having a single compressor/turbine body forces the set of rotating components to rotate at the same speed. With regard to the compressor, this is likely to create instabilities which strongly limit the maximum compression ratio. This is why the compressor/turbine body is separated into two or three mechanically independent parts (Figure 3.4).

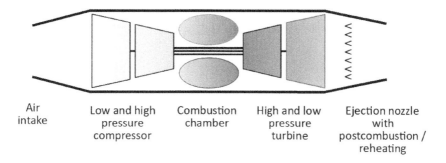

Air intake | Low and high pressure compressor | Combustion chamber | High and low pressure turbine | Ejection nozzle with postcombustion / reheating

Figure 3.4. *Configuration of a dual-spool turbojet*

We can also speak of double and triple-spool. In this case, the compression body comprises a low-pressure compression body (a booster) connected to a low-pressure turbine, and a high-pressure compression body that is then connected to a high-pressure turbine. The low- and high-pressure shafts are concentric. For the three-spool version, a compression/intermediate expansion body is placed between the high- and low-pressure bodies.

For a turboprop engine, this configuration is not optimal since it forces the propeller to rotate at the same speed as the turbine. But the propeller, which has a much greater diameter than that of the turbine, is limited in its use by the speed of the blade end, which must remain subsonic, therefore limiting the turbine efficiency which must have a lower rotation speed. For both components to optimally work together, a gearbox between the turbine and the propeller is used (Figure 3.5).

The same type of problem arises when we increase the diameter of the turbojet. Indeed, the *propulsive efficiency*, which we will discuss in the following pages, increases when the air mass flow passing through the engine increases. A larger diameter for the air inlet is therefore beneficial. The solution to this problem relies on the use of a mechanical gearbox (weight, obstruction, reliability).

The velocity difference between the air entering the engine and the air leaving it plays a decisive role on the thrust. By introducing a ducted propeller to the front of the engine, which greatly increases the speed of the gases along with the addition of a bypass channel, we obtain a dual-stream turbojet (Figure 3.6). The cold stream is deflected into the bypass and ensures a velocity difference at the outlet of the engine

(75–80% of the thrust), whereas the hot stream passes through the body of the engine (compressors, combustion chambers, turbines) and is largely responsible for powering the rotating parts. The remaining energy is then used to ensure an additional increase in the ejection nozzle output.

The ratio of the cold and hot mass flow streams is called the *bypass ratio*, with a large bypass ratio corresponding to a greater propulsive efficiency. The cold and hot streams can then be recombined inside the engine at the ejection nozzle.

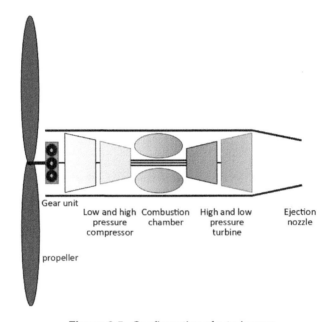

Figure 3.5. *Configuration of a turboprop*

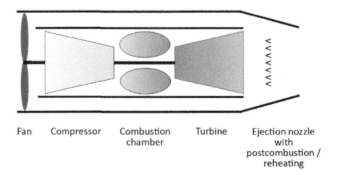

Figure 3.6. *Configuration of a dual-flow turbojet*

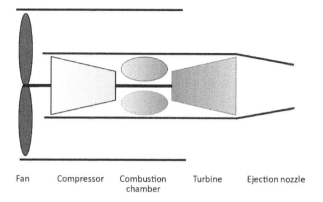

Fan Compressor Combustion Turbine Ejection nozzle
                 chamber

Figure 3.7. *Configuration of a dual-flow turbojet with a large bypass ratio*

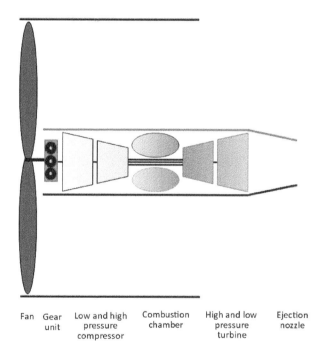

Fan Gear Low and high Combustion High and low Ejection
     unit pressure   chamber   pressure   nozzle
           compressor   turbine

Figure 3.8. *Configuration of a turbojet equipped with a geared turbofan*

The last type of improvement is the addition of a sufficiently lightweight, robust and reliable compact gearbox, which decouples the low-pressure turbine from the fan and hence ensures the optimal operation for each component. This type of turbojet is called a *geared turbofan*.

Conventional aircraft (excluding fighter jets and reconnaissance aircraft) equipped with turbojet engines generally fly at cruising speeds of altitudes between 30,000 ft and 40,000 ft (9,200–12,200 m). Indeed, the lower density air at the higher altitudes means a lower air resistance which, at the cruising speeds for steady flight, means there is a reduction in the fuel consumption. However, since the lift is also a function of the air density, for a constant mass, the lower air density must be offset by an increase in velocity in order to provide the same lift. In addition to the previous two effects, at high altitudes, the lower air density is inadequate to guarantee the combustion of the fuel. Turbojets therefore have a physical operational limit due to this altitude. To overcome this difficulty, an oxidizer must be provided to oxidize the fuel; for spacecraft which are chemically propelled, their engines come in two main types.

3.2.1.1. *Solid-propellant engines*

These engines use a propellant (Figure 3.9), i.e. a mixture consisting of a fuel, an oxidizer, a binding agent and a catalyst. The combustion of the propellant provides the hot gases which are ejected from the nozzle outlet at high speeds, producing thrust. An ammonium perchlorate mixture (oxidizing agent) along with aluminum particles (fuel) are frequently used. The geometry of the propellant blocks defines the combustion surface, giving rise to the desired thrust. The strong thrust produced by solid-propellant engines makes them particularly suitable for the propulsion of rockets during the take-off phase ("boosters").

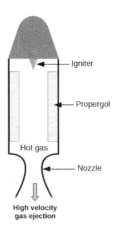

Figure 3.9. *Solid-propellant rocket*

3.2.1.2. *Liquid-propellant rocket fuel engines*

The engine, shown in Figure 3.10, consists of tanks containing the liquid propellant (fuel and oxidizing agent), injected under a strong pressure into the main combustion chamber using turbopumps driven by a turbine. The hot gases produced in the combustion chamber are expanded until it reaches a large enough speed to provide propulsion. The turbine is put into motion by the hot gases produced in an auxiliary combustion chamber.

Among the most frequently used liquid bipropellants are the cryogenic bipropellant $LO_2 - LH_2$ (liquid oxygen–liquid hydrogen), as well as the bipropellants O_2 – kerosene, N_2O_4 (nitrogen peroxide) – MMH (monomethylhydrazine) and N_2O_4/N_2H_4 (50% dinitrogen tetroxide/hydrazine) – $UDMH$ (50% unsymmetrical dimethylhydrazine).

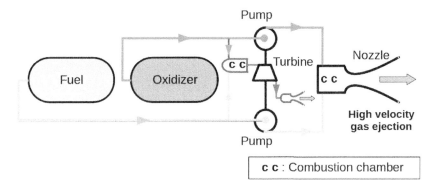

Figure 3.10. *Liquid-propellant rocket fuel engine*

The *specific impulse* (Isp) is generally used to compare rocket engine efficiency, which is defined by the relationship:

$$Isp = \frac{F}{\dot{m}\, g_0} \qquad [3.1]$$

where F is the thrust provided by the engine, $\dot{m}$ is the mass flow of propellant, and g_0 is the acceleration due to gravity at sea level.

3.2.2. **Study of the forces applied on the aircraft system during steady flight**

Before focusing on the engine engineering, we should briefly cover the overall system of an aircraft (Milne-Thomson 1958; Abbott and von Doenhoff 1959; Giovannini and Airiau 2016). In a steady horizontal flight, the aircraft is subject to four compensating forces (Figure 3.11).

3.2.2.1. *Lift*

Lift is a force that comes from the static pressure field around the aircraft and mainly around the wings and the ailerons, also known as airfoils. Note that the fuselage also plays a role in the overall lift of the aircraft. The airfoils have a camber that forces the airflow above the wing (*the extrados*) to accelerate, relative to the airflow underneath the wing (*the intrados*). This results in a difference in pressure between the intrados and the extrados of the wing which creates a force normal to the incident velocity applied to the center of pressure. At a constant velocity, increasing the angle of attack of the wing (the angle between the wing chord and the horizontal) has the effect of increasing the lift, by enlarging the "wetted" surface for the fluid and therefore also the lift, up to a certain angle ($\approx 15°$ for the typical profile). Beyond this, the lift sharply falls, a phenomenon that is called a *stall*. It should be noted that this phenomenon concerns all forms of fixed profiles (wings, ailerons), as well as rotating profiles such as propellers with a fixed or variable pitch, helicopter blades, as well as turbomachine blades. It is directly linked to compressor surge (operating instability producing pressure and flow fluctuations).

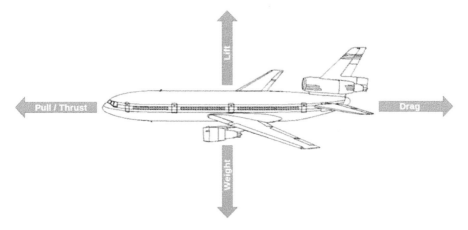

Figure 3.11. *Force assessment for an aircraft*

The lift is expressed in the form:

$$F_{Lift} = C_L \, S_{wing} \, \rho \, \frac{V_{air}^2}{2} \qquad [3.2]$$

F_{Lift}: lift [N], C_L: lift coefficient [-], ρ: air density [kg / m^3], V_{air}: air velocity [m/s], S_{wing}: surface area of the wing [m^2].

For a fixed incidence angle, increasing the velocity has the immediate effect of increasing the lift, which has to be offset by decreasing the wet surface (decreasing the incidence angle) in order to maintain horizontal flight.

The lift coefficient C_L is usually characterized by wind tunnel tests and/or computer simulations (CFD: Computational Fluid Dynamics).

A higher air density benefits the lift and explains, among other things, that for a given aircraft, why shorter take-offs are easier in colder conditions at sea level, as opposed to warm conditions and at higher altitudes.

3.2.2.2. Weight

The weight of the aircraft is a force, due to the Earth's gravitational field, based at its center of gravity and is directed towards the center of the Earth:

$$P = m\,g \qquad [3.3]$$

P: weight [N], m: mass [kg], g: acceleration due to gravity [9.81 m · s^{-2}].

During steady flight, the lift counteracts the weight. The weight of the aircraft varies throughout the flight, since fuel consumption causes the center of gravity to shift. On civil and military transport aircraft, a fuel redistribution device is used, which uses pumps in various tanks.

3.2.2.3. Drag

Drag is a force which opposes the direction of aircraft velocity, based at its center of drag:

$$F_{Drag} = C_D\,S_{wing}\,\rho\,\frac{V_{air}^2}{2} \qquad [3.4]$$

F_{Drag}: drag [N], C_D: drag coefficient [-], ρ: air density [kg/m^3], V_{air}: air velocity [m/s], S_{wing}: surface area of the wing [m^2].

There is a distinction to be made between the induced drag linked to the angle of attack of the airplane, and the parasitic drag linked to the general shape of the airplane as well as the air viscosity. Parasitic drag is the major contributor to the overall drag.

We see that the same parameters come into play which determine the lift (desired) and the drag (experienced). When designing the wing profile, the best compromise is sought between the lift and the drag forces.

An increase in the air density results in an increase in its kinematic viscosity ν ($\nu = \mu/\rho$; μ: dynamic viscosity) as well as its viscous effects, which overall increases the losses due to friction. This is one of the reasons why cruising takes place at a high altitude (33,000 ft or 10,000 m).

Similar to the lift coefficient, the drag coefficient is also characterized by tests, or from numerical simulations for a given aircraft profile/configuration.

For a given wing profile (and therefore for fixed geometric parameters), the lift and drag coefficients are solely a function of the angle of attack of the wing (α: angle between the air velocity upstream from the wing and beyond the chord of the wing), so that the ratio of the lift and drag forces is purely a function of the angle of incidence:

$$\frac{F_{Lift}}{F_{Drag}} = \frac{C_L}{C_D} = f(\alpha) \qquad [3.5]$$

This ratio is called the *lift-to-drag ratio*, and corresponds to the ability of the aircraft to cover a given horizontal distance from a predefined height. The greater the lift-to-drag ratio, the lower the thrust that the aircraft needs in order to counter the drag effects. To give an order of magnitude, current gliders can have lift-to-drag ratios of around 60, transport aircraft can have ratios between 16 and 18 and fighter jets can have one of about 10.

The *polar curve* of a profile $C_L = f(C_D)$ (Figure 3.12) illustrates the relationship between the lift and drag coefficients. At the optimum (with the tangent to the curve passing through 0), it is called the maximal lift-to-drag ratio when the lift-to-drag ratio is maximal. The end point of the curve, called the stall point, represents the point for which the flow above the upper surface of the wing no longer "sticks" to the wing profile (where the transition from a laminar flow to a turbulent flow occurs).

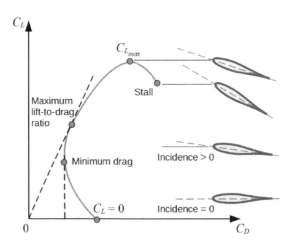

Figure 3.12. *Aerodynamic operation of an airplane wing*

3.2.2.4. *Thrust/pull*

The engine thrust/pull is a force which acts along the centerline of the aircraft directed forwards, and is based at the center of thrust. In a balanced horizontal flight, thrust compensates for the drag.

Acceleration requires either an increase in lift or an increase in thrust.

The thrust/pull applies to the aircraft at the level of its pylons which attach the nacelle system of the engine to the wings or, if without a pylon intermediary, directly to the wings or to the fuselage/cabin, in cases where propulsion is carried out by means of multi- (ATR42, CASA 160)/mono- (Pilatus PC6) turboprops and/or motor engines, or even turbojet engines in the case of military applications (Eurofighter, Rafale).

3.2.3. Definition of the propulsion forces and specific quantities of the propulsion system

As presented in section 3.2.2, the purpose of the turbojet engine is to transmit a propulsive force to the aircraft. This propelling force is called thrust.

If we consider a turbojet engine system, the sum of the forces applied to the system corresponds to the difference in the momentum generated by the turbojet engine. During a steady flight, the sum of the momenta is equal to the sum of the forces applied to the system, which can be written as:

$$F = \frac{d(m\,V)}{dt} \qquad [3.6]$$

The sum of all the forces constitutes that of a propulsive force, which is called the thrust (F_T), in addition to all of the pressure forces applied to the system:

$$F_T + A_1\,(p_a - p_1) = F \qquad [3.7]$$

where A_1 is the turbojet gas outlet sectional area, p_1 is the outlet section pressure and p_a is the ambient pressure.

In the typical case where the outlet pressure is equal to the ambient pressure, the equation simplifies and the pressure terms are eliminated.

The thrust force therefore depends on the air mass (m) entering into the engine, as well as the velocity increase that is provided by the engine to this air mass.

If we consider an aircraft moving at a velocity V_0, corresponding to the air inlet speed in the turbojet engine, and V_j the air jet velocity at the outlet of the engine, then the thrust force can be written as:

$$F_T = \dot{m}\,(V_j - V_0) \qquad [3.8]$$

$\dot{m}$: air mass flow rate.

The role of the turbojet is therefore to transfer energy to the fluid which leads to an increase in speed. This creation of energy is carried out by the combustion of a fuel, and is divided into two phases: the first consists of forming calorific energy provided by the fuel, and the second consists of transforming this calorific energy into kinetic energy.

The calorific energy addition rate corresponds to the calorific value of the fuel used (with mass flow rate $\dot{m}_f$), and we will denote it by:

$$\dot{m}_f \, HV_f \tag{3.9}$$

The rate of the kinetic energy added can be written as:

$$\frac{1}{2} \dot{m} \left(V_j^2 - V_0^2 \right) \tag{3.10}$$

Figure 3.13 summarizes the various phenomena which give rise to the generation of thrust.

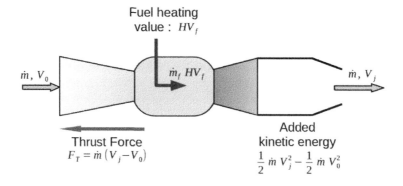

Figure 3.13. *Thermodynamic system of the turbojet*

Given that there are several different levels involved in the energy generation, it is therefore necessary to define several outputs:

– The *thermal efficiency*, which is determined by the calorific energy quantity transformed into kinetic energy, and therefore contributes towards thrust generation. It can be written as:

$$\eta_{th} = \frac{\frac{1}{2} \dot{m} \left(V_j^2 - V_0^2 \right)}{\dot{m}_f \, HV_f} \tag{3.11}$$

The thermal efficiency measures the ability to transform the energy contained within the fuel into kinetic energy.

– The *propulsive efficiency* is determined by the kinetic energy of the gases actually used to propel the aircraft, and is written as the ratio between the propulsive power and the addition of kinetic energy. The propulsive power is given by the thrust multiplied by the velocity:

$$F_T V_0 = \dot{m} (V_j - V_0) V_0 \qquad [3.12]$$

$$\eta_{prop} = \frac{\dot{m} (V_j - V_0) V_0}{\frac{1}{2} \dot{m} (V_j^2 - V_0^2)} = \frac{2}{1 + \frac{V_j}{V_0}} \qquad [3.13]$$

The propulsive efficiency characterizes the ability to transform kinetic energy into thrust.

All of these different factors give rise to the notion of an *overall efficiency*, corresponding to the amount of energy used to generate thrust, compared to the amount of energy available in the fuel:

$$\eta = \eta_{th}\, \eta_{prop} = \frac{F_T V_0}{\dot{m}_f\, HV_f} \qquad [3.14]$$

It is therefore natural to define a quantity that characterizes the quantity of fuel required to provide a specific thrust, which is called the *thrust-specific fuel consumption* (TSFC). This quantity allows us to compare the efficiencies for different engines:

$$TSFC = \frac{fuel\ consumption}{thrust} = \frac{\dot{m}_f}{F_T} \qquad [3.15]$$

The engine efficiency of a turbojet can also be defined as the amount of energy used to propel the aircraft, divided by the total amount of energy consumed. Propulsive energy is defined as the thrust force multiplied by the velocity V_0. Therefore, the efficiency can be directly related to the thrust-specific fuel consumption by the relationship:

$$\eta = \frac{F_T V_0}{\dot{m}_f\, HV_f} = \frac{V_0}{TSFC \cdot HV_f} \qquad [3.16]$$

A quick comparison of a few turbojet engines used in aviation indicates an overall decrease in the $TSFC$ over the past 50 years (Figure 3.14). It also appears that low bypass ratio turbojets, typically military engines (EJ200), for example, have a higher $TSFC$ than high bypass ratio turbojets. Modern engines with a very high bypass ratio, such as gear-driven turbofans (PW1000G), show an improvement of approximately 17% over reference turbofans such as the CFM56 or V2500.

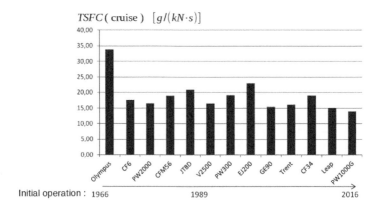

Figure 3.14. *Evolution of the TSFC*

3.3. Operating cycle analysis

3.3.1. *Hypotheses and limits of validity*

The major factors that invariably constitute turbomachines, as well as the various other configurations possibly available, have been explained previously.

Now, it is a question of studying how a turbomachine operates: the turbojet or simple-cycle single-spool turbojet engine.

This configuration was the first to historically be realized in its sub-configurations with an axial compressor in the case of the English version (Frank Whittle), and that with a centrifugal compressor in the case of the German version (Pabst von Ohain).

This relatively simple configuration, compared to other current configurations (Figure 3.3), and which is also called *simple-cycle single spool*, consists of:

– an air inlet;
– a compressor (axial or centrifugal);
– a combustion chamber;
– a turbine linked to the compressor;
– a hot gas ejection nozzle.

When used for military purposes, the exhaust nozzle can be further equipped with either a *reheat component or afterburner*, the principle of which will be studied later.

Before dedicating ourselves to the explanation of how a turbojet operates, it is advisable to carry out some simplifications. Therefore, our study will initially focus on the ideal operating cycle, which constitutes the basis of our understanding. The transition from the ideal case to the real case will then be presented later.

3.3.1.1. *First hypothesis: ideal gas model*

For this, we consider the fluid to be an ideal gas. Assuming that it has a sufficiently low pressure, and therefore the gas molecules have a sufficiently large intermolecular separation relative to the size of the molecules, then the intermolecular electrostatic forces can be neglected. Under such conditions, the thermodynamic behavior of the gas is governed by the so-called specific ideal gas equation:

$$p\,v = r\,T \qquad [1.40]$$

p [Pa]: pressure, v [m^3/kg]: specific volume, T [K]: temperature, r [J/(kg · K)]: specific ideal gas constant.

Conditions of the fluid (pressure, temperature).

Ambient air is mainly composed of nitrogen (78.08%) and oxygen (20.95 %), with the last remaining 1% coming from rare gases such as neon, krypton, xenon and helium. In the lower layers of the atmosphere, water vapor, carbon dioxide, nitrous oxide, methane, traces of dihydrogen, ozone, radon and various aerosols (dust, microorganisms), as well as other polluting gases and particles, are also present.

If we disregard the rare species and the pollutants (including water vapor), then air can be considered as a mixture of ideal gases (N_2 and O_2), whose average molar mass is about $28.965338 \cdot 10^{-3}$ kg · mol^{-1}. The ideal gas constant for this fluid mixture is equal to $r = R/M = 287$ J · kg^{-1} · K^{-1}.

The simplifying low-pressure hypothesis can sometimes become invalid, in particular when the fluid passes through the compressor.

3.3.1.2. *Second hypothesis: the composition of the fluid does not change*

The composition of the fluid is assumed to be constant throughout the entire cycle, which amounts to neglecting the composition change in the fluid as it passes through the combustion chamber, which would otherwise introduce: carbon dioxide (CO_2), water vapor and residual products from the combustion such as soot and nitrogen oxides (NO_x).

3.3.1.3. *Third hypothesis: isentropic compression and expansion*

The compression (compressor) and expansion (turbine) processes occur adiabatically, i.e. there is no heat transfer between the fluid and the exterior, and they are also quasistatic, meaning that the transformation from an equilibrium state of the fluid at any given moment to an equilibrium state at a later moment evolves infinitesimally via infinitely many intermediate equilibrium states. These processes are assumed to be *adiabatic and reversible*, and therefore isentropic (a process that evolves with a fixed entropy). In the real case, the fluid exchanges heat with the components of the turbomachine; it is therefore not adiabatic. Moreover, the fluid viscosity induces frictional forces that cause a significant amount of energy to dissipate, and so the actual process is not reversible.

3.3.1.4. *Fourth hypothesis: frictional losses are negligible*

Combustion takes place at a constant or isobaric pressure. In fact, measurements show that the loss of pressure in the combustion chamber, albeit non-zero, is relatively low. Nevertheless, significant efforts are made during the design and optimization phases of the combustion chamber to get as close to an isobaric combustion process as possible, and thus to the ideal operating cycle.

The air inlet, as well as the gas exhaust nozzle, are also considered to be free from pressure losses and heat exchanges.

The transfer of mechanical energy between the compressor and the turbine is further idealized, in the sense that we assume there are no frictional mechanical losses.

3.3.2. *Presentation of engine stations (SAE ARP 755 STANDARD)*

As we have seen in previous sections, the development process of an aeronautical engine is nowadays no longer the work of a single company, but rather of a very dense network of primary contractors along with multinational sub-contractors. In such a context, in order to facilitate the collaboration between the different parties, the English language has gradually been imposed as the lingua franca. However, this is still not quite enough to regulate the exchange of technical information in such a way as to ensure that all the collaborators speak the same "engineering" language.

First published in the 1960s and revised many times over the course of technological progress, the standard SAE Aerospace Recommended Practice 755 (SAE ARP 755) of the American company SAE International (SAE Aerospace 2013) attempts to clarify and unify, and consequently standardize, the engineering nomenclature regarding all turbine engines: the standard presents a simple nomenclatural system for the different positions passed by the fluid as it travels through the engine. *Stations* refer to fixed reference positions for which knowledge of the characteristics of the fluid medium (pressure, temperature, speed) is necessary, in order to determine the engine performance.

It should be noted that all the configurations previously discussed are covered by the ARP 755 standard, and that this standard also applies to engines derived from aeronautical turbomachines: industrial gas turbines.

Finally, the ARP 755 standard defines each evolution undergone by the fluid consistently, independent of the engine cycle type. The five transformations can be described by:

1) kinetic compression (air intake, diffuser);

2) mechanical compression, work added, fluid compression (compressor, propeller);

3) adding or exchanging heat (combustion chamber, heat exchanger);

4) mechanical expansion, work extraction (turbine);

5) mixing of fluids (mixer, exhaust nozzle).

For any engine configuration, there will always be reference stations whose definition remains unchanged (Figure 3.15). Hence, we will always find:

1) upstream air conditions (at infinity);

2) air conditions at the first reference station, located either at the entrance to the propulsion system or at an external/internal interface;

3) air conditions at the compressor inlet;

4) air conditions at the compressor outlet/combustion chamber inlet;

5) air conditions at the combustion chamber outlet/turbine inlet;

6) air conditions at the turbine outlet;

7) air conditions at the mixer inlet, (or reheat/afterburner if applicable);

8) air conditions at the exhaust nozzle inlet;

9) air conditions at the narrowest nozzle section (the throat);

10) air conditions at the exhaust nozzle outlet.

All of the positions for each base station are shown in Figure 3.15 for a simple-cycle single-spool turbojet.

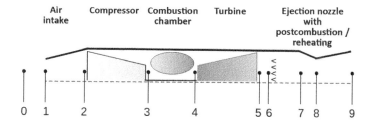

Figure 3.15. *Positions of the base engine stations for a simple-cycle single-spool turbojet*

From these base stations, it will be necessary to define new stations if we are to describe more complex configurations. To do this, the standard recommends the use of alphabetic (A, B, ...) or numeric characters (1, 2, 3, ...), separated (or not) by a special character, such as the period ".", the dash "-", the underscore "_" or the dollar "$" for clarity. Thus, for example, a station located between the high-pressure turbine outlet and the low-pressure turbine inlet would be named 45, 4.5 or even 4-5.

For the rest of this book, we will use the simplest notation without the use of special characters.

For a configuration such as the twin spool turbofan (Figure 3.16), for which there is a base as well as a high-pressure compressor along with a low- and high-pressure turbine, in addition to a bypass channel, it becomes necessary to define a station located at 25 between the low and high pressures, as well as a station located at 45 between the high- and low-pressure modules of the turbine. The same applies to stations 11 in the air supply line, 12 and 13 at the fan inlet and outlet, as well as 17, 18 and 19 along the inlet, the throat and the outlet of the fan nozzle, respectively.

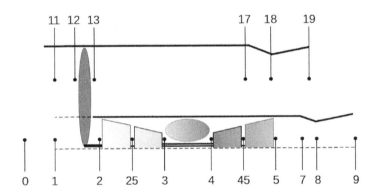

Figure 3.16. *Positions of the engine stations for a twin spool turbofan (dual flow)*

The ARP 755 standard also defines the nomenclature used, when defining the state variables of the fluid throughout its passage through the engine, of which the most common are:

– P: total pressure;
– T: total temperature;
– W: mass flow rate;
– V: velocity;
– S: static conditions;
– A: air (this is only used when there is an ambiguity in the fluid used);

– F: fuel;

– AMB: ambient conditions.

So when we say P_2, we mean the total pressure at the compressor inlet, $PAMB$ designates the total pressure conditions upstream at infinity, and WF designates the fuel mass flow rate, as opposed to $WA4$ which means the air mass flow rate at the combustion chamber outlet (including the combustion products $WA4 = WF + W_3$).

For the remainder of this chapter, we will use the nomenclature defined by the ARP 755 standard to illustrate the cycles common to all aeronautical turbomachines.

3.3.3. *Study of thermodynamic transformations and their representations in T – s diagrams*

A temperature–specific entropy diagram ($T - s$) is a diagram commonly used to visualize the thermodynamic effects of processes along an operating cycle. It should be noted that the turbomachine functions as an open cycle (continuous air intake), unlike the piston engine which functions as a closed cycle (air intake undertaken in a finite time).

In the case of ideal gas, by integrating equation [1.53]:

$$ds = c_v \frac{dT}{T} + r \frac{dv}{v} = c_p \frac{dT}{T} - r \frac{dp}{p} \qquad [1.53]$$

between the considered state (p, v, T) and a state chosen as the reference state (with index "0"), the specific entropy is expressed in the form:

$$s - s_0 = c_v \ln\left(\frac{T}{T_0}\right) + r \ln\left(\frac{v}{v_0}\right) = c_p \ln\left(\frac{T}{T_0}\right) - r \ln\left(\frac{p}{p_0}\right) \qquad [3.17]$$

The equations of the isobaric (constant pressure) and isochoric (constant specific volume) curves are obtained from relation [1.53].

– Isobaric curve:

$$T = K \exp\left(\frac{s}{c_p}\right) \qquad [3.18]$$

– Isochoric equation:

$$T = K' \exp\left(\frac{s}{c_v}\right) \qquad [3.19]$$

The curves representing the isobars and the isochores in the $T - s$ diagram are therefore exponential (Figure 3.17). According to relation [3.17], the isobaric curves that correspond to different pressure values are obtained from one another by a translation parallel to the entropy axis. The same is true for isochoric curves.

According to relation [1.53]:

$$\frac{\left(\frac{\partial T}{\partial s}\right)_v}{\left(\frac{\partial T}{\partial s}\right)_p} = \frac{c_p}{c_v} = \gamma \qquad [3.20]$$

For a given state, the gradient of the isochoric curve is greater than that of the isobaric one, and the ratio between the curves corresponds to the heat capacity ratio γ.

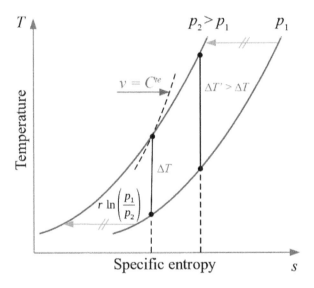

Figure 3.17. *Representation of the evolutions in the T – s diagram*

In the $T-s$ diagram, a reversible adiabatic evolution, which is therefore isentropic, is represented by a vertical line segment, parallel to the temperature axis. As shown in Figure 3.17, due to the concavity of the isobaric curves, the temperature variation, and hence also the enthalpy variation, that corresponds to a fixed pressure ratio between the initial state and the final states of an isentropic transformation increases with the temperature. This phenomenon is often called isobaric divergence in the literature.

An isothermal transformation (constant temperature) is represented by a horizontal line segment, parallel to the entropy axis.

3.3.4. *Study of the thermodynamic cycles for a gas turbine*

The operation of a turbojet corresponds to the thermodynamic cycle of a gas turbine. Work is done to rotate a machine, and a constant airflow passes through it. The air is compressed in a compressor before being heated in a combustion chamber, and then is expanded inside of a turbine (Figure 3.18).

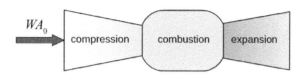

Figure 3.18. *Schematic representation of a gas turbine*

The theoretical operating cycle consists of two isentropic and isobaric curves. The heat exchanges occur at a constant pressure. This cycle therefore corresponds to the thermodynamic cycle described by George Brayton in 1872, and is commonly called the *Joule–Brayton* cycle.

Based on the representation of isentropic and isobaric processes in the $T - s$ diagram, as was developed in section 3.3.3, the two isentropic segments are connected to two isobaric ones, and we can represent the Joule–Brayton cycle in the temperature–specific entropy diagram (Figure 3.19).

The Joule–Brayton cycle is directly related to the turbomachine engine operating cycle, with the numbering used to define the cycle following the ARP 755 standard, as presented in section 3.3.2.

Thus, the phases of the ideal Joule–Brayton cycle consist of:

– a reversible adiabatic compression ($2 \rightarrow 3$):
  - the specific entropy remains constant,
  - the pressure increases from p_2 to p_3,
  - the temperature increases from T_2 to T_3,
  - the specific volume decreases from v_2 to v_3,
  - the compressor provides the fluid with the work w_{23};

– an isobaric heat input (combustion $3 \rightarrow 4$):
  - the pressure remains constant,
  - the heat quantity q_{34} is generated by combusting the fuel,
  - the temperature increases from T_3 to T_4,
  - the specific volume increases from v_3 to v_4,
  - the specific entropy increases from s_3 to s_4;

– a reversible adiabatic expansion (4 → 5):
  - the specific entropy remains constant,
  - the pressure decreases from p_4 to p_5,
  - the temperature decreases from T_4 to T_5,
  - the specific volume increases from v_4 to v_5,
  - the turbine extracts the work w_{45} from the fluid.

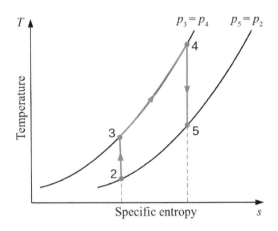

Figure 3.19. *Representation of the Joule–Brayton cycle in the T–s diagram*

The thermal efficiency of the Joule–Brayton cycle is defined by:

$$\eta = \frac{useful\ energy}{energy\ supplied} \tag{3.21}$$

The work needed to drive the compressor is extracted from the turbine shaft, with the amount available for the turbine shaft constituting the useful work.

If we neglect the variations in the kinetic and potential (gravity) energy, and given that the transformations in the compressor and the turbine are assumed to be adiabatic, then according to the first principle of thermodynamics for an open system (equation [1.5]):

$$w_{23} = h_3 - h_2 = c_p \left(T_3 - T_2 \right) > 0 \tag{3.22}$$

$$w_{45} = h_5 - h_4 = c_p \left(T_5 - T_4 \right) < 0 \tag{3.23}$$

from which we can express the useful work as:

$$w_{useful} = w_{23} + w_{45} = c_p(T_3 - T_2) - c_p(T_4 - T_5) < 0 \qquad [3.24]$$

There are no moving parts in the combustion chamber and so, as a result, it does not exchange any mechanical work with the exterior. According to equation [1.5], the amount of heat supplied from the combustion can therefore be expressed in the form:

$$q_{34} = h_4 - h_3 = c_p(T_4 - T_3) \qquad [3.25]$$

The thermal efficiency is obtained then from equations [3.24] and [3.25]:

$$\eta = \frac{|w_{useful}|}{q_{34}} = 1 - \frac{T_5 - T_2}{T_4 - T_3} \qquad [3.26]$$

With the expression for a reversible adiabatic transformation for ideal gases:

$$\frac{T_2}{T_3} = \frac{T_5}{T_4} = \left(\frac{p_2}{p_3}\right)^{\frac{\gamma-1}{\gamma}} \qquad [3.27]$$

so that the thermal efficiency can be put into the form:

$$\eta = 1 - \left(\frac{p_2}{p_3}\right)^{\frac{\gamma-1}{\gamma}} \qquad [3.28]$$

Nonetheless, the *actual cycle* of a gas turbine differs from the ideal cycle. Evolutions in the compressor and the turbine have been considered in the ideal cycle as reversible and adiabatic. In reality, however, reversibility is not guaranteed since these machines are prone to losses from friction, and cannot be considered as perfectly reversible adiabatic processes. In addition, combustion cannot be fully isobaric due to the charge losses appearing in the combustion chamber. At a constant compression and expansion rate, the compressor and turbine output temperature in a real machine will therefore be higher than those for the ideal cycle. The output pressure of the combustion chamber will hence be lower for the actual cycle than for the ideal cycle.

The actual cycle can be represented in the $T - s$ diagram by taking into account these differences (Figure 3.20).

Therefore, the energy required to provide the same amount of work will be higher in the real cycle. The energy efficiency of the actual cycle will then be lower than that of the ideal cycle.

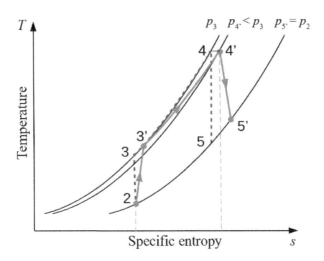

Figure 3.20. *Representation of the real Joule–Brayton cycle in the T – s diagram*

3.3.5. *Study of the thermodynamic cycle of a gas turbine, branch by branch*

In this part, we propose to study the branches of the cycle, and hence the thermodynamic components encountered successively by the air during its passage through the turbomachine. We will consider each component of the simplifying hypotheses mentioned above – a more detailed presentation of each module will be carried out in the last part of this chapter.

The theoretical operating cycle of a single-spool turbojet (Figure 3.15) is shown in Figure 3.21.

3.3.5.1. *The air intake*

The first component traversed by the airflow, the air intake conditions the airflow to ensure it passes through with a homogeneous temperature and pressure. The simplifying adiabatic flow assumption (that the air is not heated from frictional effects with the wall), reversible (without a loss of charge) and neglecting the kinetic energy of the fluid, means that the input and output of the air inlet, respectively, stations 1 and 2, become merged: $p_1 = p_2 = p_0$ and $T_1 = T_2 = T_0$. The points 0, 1 and 2 are combined in the $T - s$ diagram.

3.3.5.2. *The compressor*

The role of the compressor is to increase the energy level of the air. The compression is assumed to be adiabatic (no heat exchange) and reversible (without a loss of charge), and therefore isentropic. Assuming a constant flow (no bleed air from the compressor) and, in the absence of any mechanical losses, the specific work done by the compressor between stations 2 and 3 (temperature increase from T_2 to T_3) is obtained from the expansion in the turbine between stations 4 and 5.

The specific work conveyed to the fluid by the compressor (ideal gas hypothesis) is expressed by:

$$w_{23} = c_p \left(T_3 - T_2 \right) \quad\quad\quad [3.29]$$

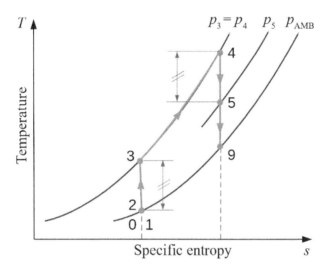

Figure 3.21. *Representation of the theoretical cycle of a simple-cycle turbojet engine*

3.3.5.3. *The combustion chamber*

Once the pressure and temperature p_3 and T_3 levels have been reached, the air passes into the combustion chamber where the fuel (a much lower quantity than that of air) will be added to it, before the oxidizer/fuel mixture is ignited. Thus, the energy level of the air is increased by the addition of heat and by extracting the chemical energy contained within the fuel. The combustion process is extremely complex and will be covered in more detail later. Modern combustion chambers are characterized by pressure losses of the order of 2–3%. Under our simplifying assumptions, these losses are neglected, so we will speak of combustion at a constant pressure. The passage through the combustion chamber therefore takes place at a constant pressure, where only the temperature changes from T_3 to T_4 along the isobar p_3.

The quantity of heat that the fluid receives in the combustion chamber is:

$$q_{34} = c_p \left(T_4 - T_3\right) \qquad [3.30]$$

3.3.5.4. *The turbine*

The purpose of the turbine is to convert the energy of the fluid leaving the combustion chamber into mechanical energy to drive the compressor. At the turbine inlet, the fluid has the pressure $p_4 = p_3$ and temperature T_4. At the turbine outlet, the pressure is p_5 and the temperature is T_5. Mechanical losses between the turbine and the compressor are assumed to be zero. The expansion assumptions within the turbine are similar to those made for compression: expansion is assumed to be reversible and adiabatic. Thus, we can write, ostensibly neglecting the different nature of the gases present in the compressor and the turbine, and the addition of fuel:

$$w_{23} = c_p \left(T_3 - T_2\right) = c_p \left(T_4 - T_5\right) = |w_{45}| \qquad [3.31]$$

or even:

$$T_3 - T_2 = T_4 - T_5 \qquad [3.32]$$

3.3.5.5. *The exhaust nozzle*

Similar to the air inlet, the exhaust nozzle is considered adiabatic and without any losses from friction. Its role is to convert the energy of the fluid exiting the turbine outlet in the state with T_5, p_5 and expanding it to the state T_9, p_9 at the nozzle outlet, where the gas velocity is V_9. If, at the outlet of the nozzle, we have $p_9 = p_0$, then we say that the nozzle is suitable.

Assuming that the velocity V_5 at the turbine outlet is small compared to the velocity V_9 then, in the case of a nozzle without an afterburner, by the first law of thermodynamics, we have:

$$h_5 = h_9 + \frac{V_9^2}{2} \qquad [3.33]$$

which can be written still, with the ideal gas hypothesis, as:

$$V_9 = \sqrt{2\, c_p \left(T_5 - T_9\right)} \qquad [3.34]$$

3.3.6. *Improvements to the Joule–Brayton cycle*

The efficiency of the Joule–Brayton cycle can be understood as the area between the two isentropic and the two isobaric curves. Hence, improving the cycle corresponds to increasing this area. There are two simple ways to do this, at least theoretically. We will see that these improvements come up against real physical barriers, however, which are extremely complex to overcome.

3.3.6.1. *Increasing the turbine inlet temperature*

The first way is to "push" the isentrope corresponding to the expansion in the turbine to the right. As the position of this isentrope is directly related to T_4, the turbine inlet temperature, T_4, should be increased to $T_{4'}$. If the work done by the compressor w_C remains constant (Figure 3.22), while the enthalpy drop from the expansion increases (by the isobar divergence effect – see section 3.3.3), we find that the energy remaining after extracting the work needed to drive the compressor (useful work) is much greater than that of the 4'–5' isentrope.

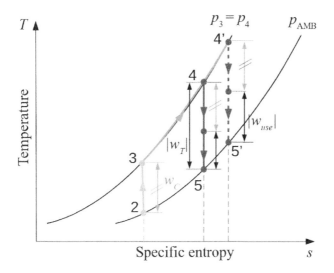

Figure 3.22. *Representation of the Joule–Brayton cycle, after increasing the turbine inlet temperature in the T – s diagram*

This energy can be converted either into kinetic energy, as to create a thrust increase by increasing the velocity differences between the engine inlet and the outlet (in the case of a turbojet), or else be converted into mechanical energy to drive a propeller (in the case of a turboprop). In the last case, it will be necessary to increase the work done by the turbine $|w_T|$ by as much as possible.

Increasing the turbine inlet temperature therefore seems to be the more desirable, and many developments are moving in this direction. It would appear that increasing the mass flow of the fuel could also work, which would increase the quantity of heat supplied to the fluid without too many problems. However, apart from the fact that such a solution is economically and ecologically undesirable, there is a much stricter physical limit to this increase: the temperature limit of the turbine blades. Indeed, as these are placed directly at the combustion chamber outlet, they would experience this heat increase directly.

Simply changing the material to allow for higher maximum temperatures is not enough. Insulating layers are therefore used, applied directly to the surface of the blades and, should this still be insufficient, cooling systems for the blades that use an internal flow of fresh air can be used from the high-pressure compressor. All of these workarounds come at a cost: a heavier weight coming from the heat shields, and a lower cycle efficiency from the cooling system. Indeed, this is because the air taken for the cooling purposes, which has undergone compression, no longer contributes in the overall cycle.

3.3.6.2. *Increasing the compression ratio of the compressor*

Another way to improve the Joule–Brayton cycle is to increase the pressure level, p_3. Hence, by shifting to a higher level isobar, the area delimited by the isentropes and isobars increases (Figure 3.23).

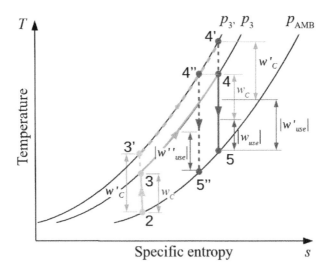

Figure 3.23. *Representation of the Joule–Brayton cycle, showing the effect of increasing the compressor outlet pressure in the T – s diagram*

It should be noted that, up to the multiplicative factor c_p, the variations in the temperature during the compressions and expansions give rise to work. Increasing the pressure level at the compressor outlet (2–3' being greater than 2–3) induces an increase in the work done by the compressor w'_C. As explained previously, the exponential form of the isobars does not allow for the 4'–5 isentrope to be preserved, because then the temperature $T_{4'}$ is much too high. It is necessary therefore to lower the attained temperature level. If we place ourselves at the temperature $T_{4''} = T_4$, the level of energy remaining after extraction of the mechanical work of the turbine to drive the compressor is reduced from $|w'_{use}|$ to $|w''_{use}|$, thus limiting the potential gain that would come from an increase in the compression ratio.

What is left now to consider is that for the compressor to generate more work from compression, each of the given compression stages must themselves provide a large enough amount of work: this is the source of operational instabilities for the compressor (surge, stall) that can be extremely harmful. Alternatively, it is possible to increase the number of compression stages, though this increases the overall weight of the engine, and therefore causes a greater performance loss in addition to increasing the complexity of the propulsion system.

Therefore in the real case, improving the Joule–Brayton cycle constitutes an extremely complex and multidisciplinary problem which can require innovative concepts that diverge from the notion of conventional gas turbine diagrams, if necessary.

3.3.7. *Thermodynamic improvements for a gas turbine using energy regeneration*

On the graph representing the Joule–Brayton cycle in the $T - s$ diagram (Figure 3.24), we note that T_5 is greater than T_3. This situation is also the case for actual engines, where the temperature at the outlet of the turbine is higher than that at the outlet of the compressor. This temperature difference represents an energy reserve, and it is interesting to use this in ways to improve the thermal efficiency of the gas turbine. For example, by using a heat exchanger, it is possible to recover some energy from the hot gases at the turbine outlet, and then use it to increase the air temperature at the compressor outlet.

Given a heat exchanger with no heat loss, by the first principle of thermodynamics, the quantity of heat transferred from the hot gases at the turbine outlet to the gases at the compressor outlet is:

$$q_r = c_p \left(T_5 - T_{5'} \right) = c_p \left(T_{3'} - T_3 \right) \qquad [3.35]$$

The maximum temperature value $T_{3'}$ likely to be attained at the exchanger outlet is equal to T_5 (in the case of an exchanger with an efficiency of 1 corresponding to an exchange surface tending towards infinity).

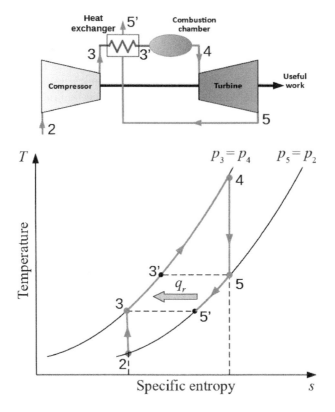

Figure 3.24. *Representation of the Joule–Brayton cycle with energy recovery in the T – s diagram*

Therefore, for the Joule–Brayton cycle with energy regeneration, its thermal efficiency can be written as:

$$\eta = \frac{|w_{use}|}{q_{3'4}} = \frac{|w_T| - w_C}{q_{3'4}} = \frac{c_p(T_4 - T_5) - c_p(T_3 - T_2)}{c_p(T_4 - T_{3'})} \qquad [3.36]$$

and, in the case of an ideal heat exchanger with an efficiency of 1 ($T_{3'} = T_5$):

$$\eta = 1 - \frac{T_3 - T_2}{T_4 - T_5} \qquad [3.37]$$

which can also be written as:

$$\eta = 1 - \frac{T_3}{T_5}\left(\frac{p_2}{p_3}\right)^{\frac{\gamma-1}{\gamma}} \qquad [3.38]$$

As the ratio T_3/T_5 is less than one, the thermal efficiency of the cycle with heat regeneration is greater than that of the Joule–Brayton cycle without any regeneration.

3.3.8. *Thermodynamic improvements for a gas turbine using staged compression and expansion*

Analysis of the $T - s$ diagram of the Joule–Brayton cycle also shows that the work required during compression increases when the air temperature increases, which follows from the exponential behavior of the isobars. The same occurs for the work generated by the turbine, as it increases when the temperature of the hot gases is greater.

It is therefore interesting to perform the compression and expansion in stages, with some extraction (cooling) or addition (combustion) of energy occurring between each stage, for example, as in the following system (Figure 3.25) with two stages.

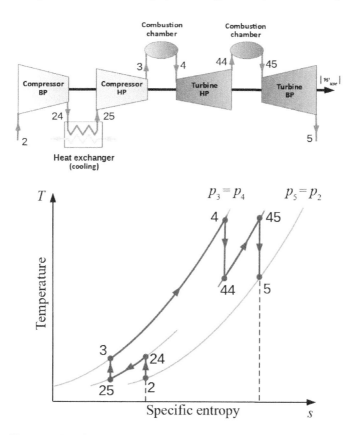

Figure 3.25. *Gas turbine with staged compression and expansion*

In the case of a turbofan engine, it is common for it to include two compressors and two separate turbines (a low-pressure branch and a high-pressure branch). It is therefore possible to include the thermodynamic elements between the two branches, without running into architectural difficulties.

For a two-stage system with a compressor and turbine, an example cycle is depicted in Figure 3.25.

Note that the outlet temperatures of the LP (T_{24}) and HP (T_3) compressors, as well as the outlet temperatures of the HP (T_{44}) and LP (T_5) turbines, are identical, subject to a suitable transition for the pressures between the LP and HP stages. In the theoretical case with infinitely many compressor and turbine stages, compression and expansion tend towards an isothermal processes. If we also include energy regeneration that was presented in section 3.3.7, the cycle becomes equivalent to the Ericsson cycle (Figure 3.26), whose thermal efficiency reaches the maximum efficiency of the Carnot cycle:

$$\eta = 1 - \frac{T_2}{T_4} = 1 - \frac{T_c}{T_h} \qquad [3.39]$$

The different stations of the Ericsson cycle cannot be standardized by the ARP 755 standard. It is not possible to represent the current classical internal combustion turbomachines using the theoretical Ericsson cycle.

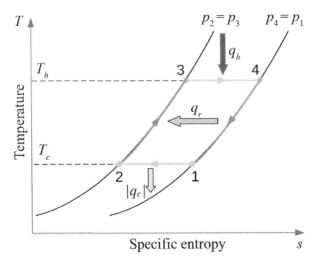

Figure 3.26. *Representation of the Ericsson cycle in the T – s diagram*

3.4. The actual engine

3.4.1. *Development cycle of the turbomachine (turbojet)*

The development cycle of an aeronautical or industrial turbomachine is generally divided into seven distinct and successive phases, numbered from 0 to 6 (Figure 3.27).

The transition from one phase to the next is subject to two forms of reviewing: one is technical/technological and the other is financial, and are called "gates". This "gating process" allows the company to ensure that, at each phase of development, the proposed technology matches the market demand, the development costs involved and the benefits that can be derived from the implementation of the product. In light of the complexity of turbomachine development, the same gating process is used at the level of each module (air intake, low- and high-pressure compressors, combustion chamber, high- and low-pressure turbines, exhaust nozzle), in such a way as to minimize technological and financial risks. Each technological and/or financial gate decides either the continuation or termination of the project.

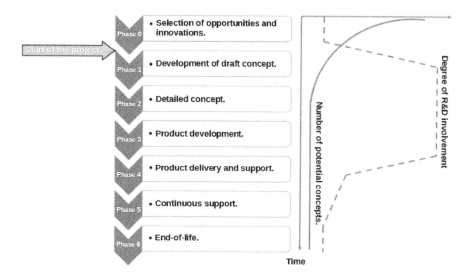

Figure 3.27. *Development phases of a turbomachine*

It should be noted that the introduction of new technologies for aeronautical equipment (aircraft and/or engine) requires a lengthy amount of work to meet the certification requirements (safety aspects), which explains the incremental pattern for innovation (small changes between two comparable products) within the aircraft industry.

3.4.1.1. *Phase 0: Selection of opportunities and innovations*

The project initiation phase focuses on the detailed study of various environmental aspects (economic situation, demand, need) which will decide whether to start or postpone the project. This phase includes the following activities:

– identification of potential markets (new or already existing);

– identification of customer needs (civilian, military, single/twin/tri/four-engine(s), weight, maximum thrust, etc.);

– development of the product and market strategy;

– identification of one's own financial and investment capacities (credit, etc.);

– development of a long-term technological and financial plan;

– introduction of the new project within the overall operation of the company.

3.4.1.2. Phase 1: development of the draft concept

This phase involves developing a sound technological and financial plan to deal with external possible constraints (market, customer demand, technology available and/or to be acquired, etc.). In this phase, the following points need to be specified:

– study of alternative markets (another aircraft, use as an industrial power generation turbomachine);

– study of alternative technological solutions/concepts;

– specification of the final cost to be achieved (target cost);

– assessment of technological and financial risks;

– implementation of a concept for the supply chain (subcontractors, rank);

– examination of possible cooperation (companies and/or universities), creation of partnerships/joint ventures;

– assessment of the investment and implementation plan.

3.4.1.3. Phase 2: detailed concept

This precision phase during the development process makes it possible to specify in more detail the technological and financial concepts. During this phase, the following aspects are examined:

– concept choices guaranteeing the customer's needs and target costs;

– implementation of a risk mitigation plan;

– product development plan;

– ensuring the availability of resources (research and development, production, editing, time, side project(s)).

3.4.1.4. Phase 3: product development

This phase includes most of the work done collaboratively between the various disciplines in the research and development centers. The first prototypes are assembled and tested in real conditions, to obtain the European (European Union Aviation Safety Agency: EASA) and American (Federal Aviation Administration: FAA) certifications necessary to receive the proper authorization for operational use. This phase includes the following activities:

– detailed analysis of the product (construction/design, performance, aerodynamics, thermal, structural);

– detailed analysis of the production and assembly of the product;

– development of the necessary tools;

– creation of the first models, prototypes and tests in real conditions (bench test on the ground, in flight, tests at maximum charge, endurance tests, foreign object damage tests, ice protection system tests);

– pilot implementation to ensure stable production series: this activity is characterized by a thorough collaboration between the R&D and production centers;

– increasing the production series in order to obtain the desired product quantity;

– creation of the engine manual, including the maintenance plans as well as the limits of use.

3.4.1.5. *Phase 4: product delivery and support*

After having obtained the type certificate, the engine actually begins to operate for the first time having been integrated into an aircraft, on behalf of the customer. This phase is characterized by:

– a production volume which can be adjusted according to demand (precise demand for engines and spare parts);

– customer support once in operation;

– integration of feedback from experiences in real conditions (operation, production, maintenance);

– integrating modifications into the certified engine (retrofit, updates, derived versions).

3.4.1.6. *Phase 5: continuous support*

This phase is characterized by a significant reduction in the number of new engines, and by an increase in maintenance activities (maintenance, repair, operation). The activity of research and development centers is also limited. Hence, this phase is characterized by:

– delivery of high-quality customer service (troubleshooting, inspection during small and large visits);

– prediction of spare parts demand as accurately as possible (logistics and strategy);

– reduction of production costs (e.g. by outsourcing and by promoting competition between subcontractors).

3.4.1.7. *Phase 6: end-of-life*

This last phase, often forgotten, relates to the end-of-life of the engine. The maximum life span of an aircraft having followed the maintenance plans (major inspection every five years) is estimated to be between 25 and 30 years, corresponding to 100,000 cycles (one cycle = take-off and landing fight). A turbomachine is planned to last the same time. However, the age of the aircraft and of the engine has negative consequences on its operational use (higher maintenance cost, decreasing passenger approval, etc.). Thus, the average age of aircraft in Europe is only 10–15 years. So, when the engine system is no longer competitive for a company, it can either be resold (returning to phase 5) or reprocessed. Reprocessing it is a legal obligation of the engine owner. Nowadays, recycling aspects must be taken into account when designing the engine, which greatly simplifies the reprocessing process. In addition to environmental aspects, recycling has significant financial aspects to which airlines and engine manufacturers are becoming increasingly sensitive.

There are several levels of reprocessing:

– the highest level is that of the engine system itself if it is yet to reach its end-of-life. In this case, it can be installed on another aircraft having a similar sort of engine after inspection;

– the second best level is on the modular level; modules (turbine, LPC, HPC, combustion chamber, HPT, LPT, ancillary systems, etc.) with the potential to be reused;

– the next level is the reprocessing of spare parts, the market for which is in full expansion. After re-certifying the parts, this allows the company to either use them in engines of the same type, or to make a financial profit by reselling them;

– the lowest level is that involving the raw metal. Many different materials (steel, aluminum, titanium, etc.) and exotic materials (special alloys) are present in the engine. Once sorted, these materials can be resold. It should be noted, however, that quality control criteria for aeronautical materials do not allow for the use of recycled materials. Hence, it is not possible to melt down a turbine blade to create a new one, for example.

3.4.2. *Technical disciplines in development*

Many technical disciplines come together during the detailed concept and engine development phases. Each one of these disciplines is responsible for a field of study, and clearly defined responsibilities and interactions.

3.4.2.1. *Pre-design*

The role of pre-design is to determine the engine architecture(s) that will meet the customer specifications (thrust, speed, size). There are numerous pre-design parameters:

– fan diameter;
– dilution ratio;
– number of compressors (1, 2, 3);
– number of compression stages per compressor, and the compression ratio;
– type of combustion chamber;
– number of turbines (1, 2, 3);
– number of expansion stages per turbine;

– exhaust nozzle architecture: convergent, convergent–divergent, afterburner, mixer.

The search for the optimal configuration is based on the thermodynamic considerations of the Joule–Brayton cycle. For this, 1D tools – often developed within the company itself – are used. Note that there are also commercial tools such as GasTurb^{TM}.

3.4.2.2. *Performance*

Once the overall architecture is fixed, it is necessary to refine our knowledge of its performance. For this, the same 1D tools can be used to model each of the engine modules as well as their interactions more precisely. The goal is to obtain a global 1D vision thanks to the knowledge of the physical quantities $(p, T, \dot{m})$ for each engine station. To do this, each module is modeled within a 1D network in which the nodes correspond to its inputs and outputs (and therefore to each engine station), and the lines correspond to the average airflow within the module itself and is characterized by its losses (thermal and aerodynamic efficiency). At each node, the mass flow and energy conservation equations are solved, as well as the momentum conservation equation for each module. The resulting nonlinear system of equations is solved iteratively with respect to its design purpose (e.g. continuous cruise for civil use), in addition to the other important operational points that will constitute the flight envelope (deceleration, taxiing, takeoff, initial climb, cruising, etc.).

At the end of this analysis, each point that constitutes the operational cycle of the engine is thermodynamically characterized, and the general desired qualities for each of the modules are set (the target efficiency).

The following analysis consists – for the most part – of obtaining from the general concept of the architecture and performance, all of the engine components that will make it possible to move from the intangible concept to a reality. Initially, all the dimensions of the parts are determined using numerical simulations.

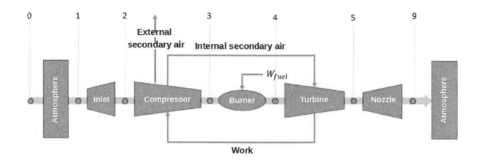

Figure 3.28. *"Performance" model of a simple-cycle single-spool turbojet*

3.4.2.3. *Primary airflow*

The first parts to be designed in the engine are the blades. Indeed, their configuration is directly related to the quantities previously calculated by the performance.

3.4.2.4. *Aerodynamics*

Aerodynamics focuses on designing the blades (number, profile) for each compressor and turbine stage, so as to obtain the desired quantities calculated for the performance (efficiency). The methods implemented to achieve this are specific to the discipline, and include CAD tools (for blade design) coupled with 2D and 3D computational fluid mechanics methods (CFD) (stationary and non-stationary Navier–Stokes equations).

3.4.2.5. *Construction/structure/thermal*

The design of the blades is carried out in close collaboration with the construction (CAD), thermal (finite element modeling in 3D) and structural (also finite element modeling in 3D) disciplines, to ensure that the life span of the blades is sufficient (polycyclic and megacyclic fatigue) and satisfies the specifications. Spectral analysis (vibration, resonance, aeroelasticity) is carried out to avoid any harmful vibratory phenomenon as well as creep calculations (delayed irreversible deformations of a material subjected to a constant stress, lower than the elastic limit of the material for a sufficiently long period, in our case centrifugal force and heat) taking place to predict the deformation behavior of the blades.

This process is in no way linear, and gives rise to several iterations between the various disciplines before finally determining the airfoil dimensions and then freezing the main flow channel design (form of the channel, number of stages, airfoil form, etc.).

With the turbomachine components defined, it is then necessary to integrate them together to construct an engine.

For example, all the rotating pieces must be installed on bearings, which will need to be lubricated. It will be necessary therefore to integrate a lubrication system (as a closed circuit), which will have the task of lubricating the bearings while also keeping the oil at a suitable temperature. Hence, it is also necessary to cool the lubricating liquid by adding heat exchangers to the fuel system, which will then improve the vaporization and therefore the combustion ...

These same rotating pieces possess extremely high energy (of the order of magnitude for a high-pressure compressor going at the 20,000 rotations/minute at take-off speed; a blade of about 50 g whose center of mass is located 20 cm from the rotational axis has an energy of 400 kJ). In the case of breakage at the base of blade, the blade must be contained (a certification) inside of the engine to avoid damage to the aircraft, or worse of a passenger or of the ground crew. It is necessary therefore to design an engine casing that meets such requirements, without penalizing the engine performance by an excessive increase in mass (such as with a thicker housing).

These two examples illustrate the extreme complexity, as well as the multi-disciplinary nature of the engine system design and its certification.

3.4.3. *Some specific problems of each module*

In the following sections, we will consider a module-by-module analysis of the simple-cycle single-spool turbojet during the postcombustion. The term "module" is reserved for the different thermodynamic elements encountered by the air during its passage through the turbomachine. These modules are comprised of:
– the air intake;
– the compressor;
– the diffuser;
– the combustion chamber;
– turbine;
– the ejection nozzle.

We will proceed as indicated in the list, by presenting the modules in the physical order that they are encountered.

3.4.3.1. *The air intake*

The air inlet at the front of the turbojet provides the rest of the turbomachine with the air mass flow required to generate the thrust under all of the operating conditions of the engine. At first glance, this element looks like a circular or elliptical sectional flow channel (subsonic civil A320, sub/supersonic military Mirage 2000, Rafale), or even rectangular (civil and military sub/supersonic Concorde, Jaguar, Eurofighter). Although simple in appearance, it requires a very particular care since it is important, given that its role is decisive for the rest of the turbomachine. This module constitutes an interface between the engine manufacturer, the nacelle manufacturer and the plane manufacturer: the three parties must work closely together to meet the needs of turbomachine and the aircraft to ensure an optimal installation. The most common installation on medium- or long-haul aircraft (civilian or military) is when the engine is connected to the aircraft via a mast called the pylon or, for short- and medium-haul aircraft, towards the rear of the fuselage. This configuration isolates the engine from the harmful effects of the wings and fuselage by as much as possible, but maximizes the resistance that the aircraft experiences (large front surface). For fighter jets and bombers flying in their trans- and supersonic fields of operation, the engine(s) are installed inside the aircraft, and the air inputs are installed upstream at the junction between the wings and the fuselage.

In subsonic (civil aircraft), transonic (business jets) or supersonic (fighter jets), the air intake and the supply channel that follows it should be designed as to ensure that the air mass flow required by the turbomachine is properly delivered, and moreover that the flow arriving at the compressor inlet (station 2) is the most homogeneous, stable and generates as little losses as possible. These conditions must be satisfied throughout the entire operating range of the turbomachine, and thus of the aircraft which can include extremely varied phases.

Therefore, we will make note of the following phases:

– *Ground idle*: for example, after startup, or during the wait before entering the runway for take-off. This operating phase is characterized by a zero upstream velocity ($V_0 = 0$) as well as atmospheric pressure p_0 and temperature on the day at the airport T_0, and also the level of humidity. It should be noted that these airport conditions are subject to large variations for a given airport (seasonal and meteorological effects) as well as between airports (Table 3.1). Thus, we will talk about altitudinal pressure; we also take into account the atmospheric pressure depending on the altitude and cyclonic conditions.

– *Take-off*: characterized by a maximum engine velocity and a speed up to the order of Mach 0.3, with the same atmospheric conditions as at ground idle.

– *Cruise*: characterized by a velocity of the order of Mach 0.8. For a civil airplane, its altitude is capable of varying between 30,000 and 40,000 feet (9,200 and 12,200 meters) with temperatures close to -54°C and atmospheric pressures of the order of 300 hPa.

Airport	Altitude [m]	Mean temperature 2017 [°C]	Maximum temperature averaged 2017 [°C]	Minimum temperature averaged 2017 [°C]	Average annual rainfall 2017 [mm]	Minimum atmospheric pressure 2017 [hPa]	Maximum atmospheric pressure 2017 [hPa]
Amsterdam	14	15	18	12	670	994,4	1039,3
Dubai	0	30	35	25	36	994,1	1026,5
Lapaz	3642	8	15	0	576	1032	1041

Table 3.1. *Average atmospheric data for three airports*

In addition, there are wind effects (force and direction), which are more important when the aircraft velocity is low, as well as the effects coming from the installation, such as the distance between the nacelle and the ground, or the possible interactions between the aircraft parts (fuselage, pylon, leading edge of the airfoil), and/or with other engines. The diagrams (Figures 3.29 and 3.30) illustrate the different types of flows as a function of the various flight phases, and therefore of an engine operating for a subsonic flow.

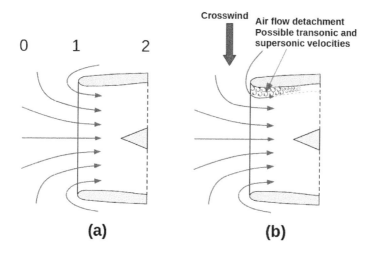

Figure 3.29. *Inlet flow without an axial velocity ($V_0 = 0\ m/s$): (a) without and (b) with crosswind effects*

When the engine is static, the air inlet takes in the amount of air required by the compressor from a very large region around the engine, of which there are regions located outside of the nacelle situated downstream from station 1. To avoid any flow detachment of the air streams in such flow configurations, the nacelle edge always has a rounded shape.

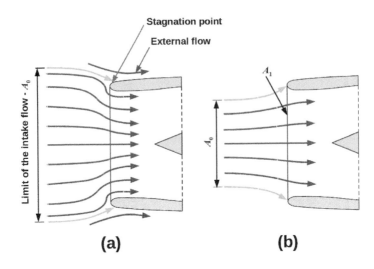

Figure 3.30. *Entry flow (a) for a low-velocity flight and (b) at cruising velocity*

However, in crosswind conditions, the nacelle edge configuration may not be sufficient to ensure a laminar flow. The flow detachment of the air streams, in the same way as the stall of a wing profile, results in a locally turbulent zone that creates one or more circumferential pressure heterogeneities in the flow which can interfere with the operation of the downstream compressor (see the compressor section).

The tailwind component is another aggravating effect for the operation of the turbomachine when the aircraft has little or no speed. Indeed, just like what happens when using thrust reversers if they are not deactivated in time, it can happen that the engine re-ingests the hot gases that it has produced itself, or from another engine in a four-engine configuration (A380, A340, A400M, B747) or, in the military case, exhaust gases from projectiles (shells, rockets, missiles). This leads to a significant increase in the air temperature that affects the density and the air mass flow rate, which disrupts the operation of the compressor and can lead to a serious degradation of the whole turbomachine (compressor surge, compressor stall, flameout by flame extinction).

Any effect that causes heterogeneity (pressure, temperature) in the ingested air is called an air inlet flow distortion. These effects, which in reality are always present to a greater or lesser extent (ground effect, wind), must be taken into account when designing the air inlet.

Increasing the aircraft velocity results in the flow lines becoming narrower, and shifts the stagnation point from the outside of the nacelle edge to the inside (Figure 3.30).

The continuity equation between stations 0 and 2 is written as:

$$\rho_0 \, V_0 \, A_0 = \rho_2 \, V_2 \, A_2 \qquad [3.40]$$

or even as:

$$A_0 = \frac{\rho_2 \, V_2 \, A_2}{\rho_0 \, V_0} \qquad [3.41]$$

If we consider the density change between stations 0 and 2 to be negligible, that A_2 is determined by the compressor design, and that V_2 depends solely on the compressor operation, then we find that A_0 is solely a function of the aircraft velocity, inversely proportional to it. For a cruising flight for which V_0 is close to Mach 0.8, the cross-sectional area A_0 becomes smaller than that of A_1. This leads to an increase in the pressure upstream from the air inlet, as well as in a decrease in velocity. Compression continues within the air inlet whose cross-section widens (diffuser) in a continuous way to transform the kinetic energy of the flow into potential energy of the pressure. This effect of the diffuser allows for an increase in the pressure level which has two beneficial effects for the compressor:

– the flow velocity at station 2 is reduced to an acceptable level;

– thanks to the pressure increase, the effectiveness of the compression phase in the Joule–Brayton cycle is improved. A substantial gain is thus obtained during the operational ranges of sub- or supersonic cruising.

To maximize this pressure gain, the air inlet must be designed so that part of the kinetic energy is not transformed into thermal energy from frictional effects. Thus, the walls which constitute the air inlet should be as smooth as possible, and without any sudden change in the section (stair-step) in the subsonic case.

In the case of supersonic flight, the air velocity in the compressor has to be reduced to subsonic conditions ($\approx$ Mach 0.4–0.5), while trying to reduce any pressure losses. To do this, the geometry of the flow channel within the air inlet is no longer fixed but can change depending on the operational velocity, and in particular the velocity of the aircraft. The reduction in the kinetic energy is done in this case by means of straight, oblique shocks, or by a combination of the two types of shocks that constitute a flow singularity. Since the theory of shock waves is not the objective of this textbook, we will consider them in a simplified way, such that:

– the total temperature is conserved (adiabatic);

– the total pressure decreases;

– the velocity decreases;

– the static quantities of pressure, temperature and specific volume increase.

The process is an immediate one and can be initiated and controlled by geometric singularities, such as an abrupt transition from a horizontal to a sloping wall, or by using the tip of a conical profile.

3.4.3.2. *The compressor*

The cycle of a gas turbine includes a compression of the air, initially at atmospheric pressure (International Standard Atmosphere – normal atmospheric pressure: 1 bar, 15°C) prior to the combustion of the fuel, and the energy released by the combustion is proportional to the mass of air consumed. To increase the efficiency of the combustion phase, it is necessary to supply air at a pressure higher than that of the atmosphere.

To achieve compression, the first method, which was developed during World War II in England, is that of a radial or centrifugal compressor. This type of compressor has the advantage of being quite compact, but delivers a relatively low pressure ratio (maximum 10:1). It is often used for auxiliary power units (APUs) and turbochargers due to its small footprint.

The second method, developed at the same time in Germany, is that of the axial compressor. This type of compressor allows for high compression ratios (60:1) by using several stages in a series; hence, it is of greater use here. Since it is mainly used for aeronautical propulsion, we will focus on explaining its operation.

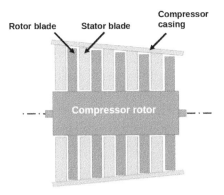

Figure 3.31. *Cross-sectional diagram of a compressor*

An axial compressor, as shown in Figure 3.31, has a convergent conical section: the volume occupied by the air in the inlet section (station 2) is greatly reduced by the work done by the compressor. The compressor comprises a series of fixed blades (stators) followed by a series of rotating blades (rotors). They are connected to the turbine by the rotor shaft, which provides the mechanical work necessary to drive the rotor part of the compressor, thereby providing the required compression work. The succession of a stator and a rotor is called a compression stage. In Figure 3.31, for example, a five-stage compressor is shown.

Figure 3.32 shows the current lines and flow velocities of a stator and rotor during a compressor stage. The stator/rotor succession is not arranged coincidentally, since the stator has two functions.

The first of these two functions is to pack the air together for the next rotor stage. Indeed, the yaws of a rotor, like the wings of an aircraft, need an optimal airflow (i.e. angle of incidence at the leading edge) to operate them most effectively. The stator therefore allows for a "rectification" of the flow, which is why it is sometimes called a rectifier.

As the axial component of the input speed velocity can vary according to the different operational phases, and therefore adjust the velocity component at the stator output, the first compressor stages have stators whose angles of incidence (timing) can vary, so as to optimally adapt the flow for the following stages.

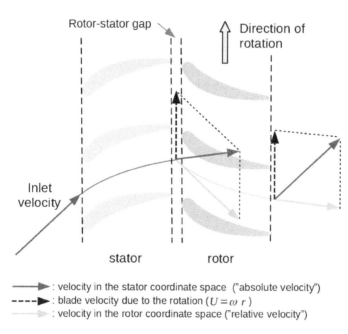

Figure 3.32. *Operational scheme of a compressor stage*

The second function of the stator is, by the virtue of its shape, to transform part of the kinetic energy into a static pressure configuration. To do this, we use a geometric phenomenon (diffuser effect, Venturi effect) that was described by Bernoulli. With the compression rate per stage remaining low, the air can be considered to be an isovolumetric fluid and, between the inlet and the outlet of the stator, Bernoulli's equation for the stream surface traversed by the fluid is written as:

$$p_1 + \rho \frac{V_1^2}{2} + \rho g z_1 = p_2 + \rho \frac{V_2^2}{2} + \rho g z_2 \qquad [3.42]$$

where ρ is the density of the fluid, V is the fluid velocity, g is the gravitational constant, z is the altitude and p is the static pressure.

As a first approximation inside of the engine, the altitude difference between two points located on the center line is negligible. This results in:

$$p_2 - p_1 = \rho \frac{V_1^2}{2} - \rho \frac{V_2^2}{2} \qquad [3.43]$$

A decrease in velocity results in an increase of static pressure. The pressure therefore increases as the stator passes due to diffusion effects.

We have seen that a flow disturbance in the air inlet can have harmful consequences on the operation of the compressor. We will see in more detail how these harmful phenomena come about.

The compressor, unlike the turbine, has an intrinsic function of opposing the natural airflow, namely from a high pressure to a low pressure. The more that the air progresses inside of the compressor, the more that its potential energy increases from the pressure, and the more difficult it is to make it progress further. This phenomenon is an instability vector.

The rotor and stator blades are similar to airplane wings. If the angle between the chord of the profile and the (incidental) fluid velocity increases beyond a certain value ($\approx 15°$), then the flow stalls (Hill and Peterson 1992). It forms a recirculation region which decreases the section of the flow. Thus, between two neighboring blades, the required air mass flow cannot be provided; the reduction in cross-sectional area causes an increase in velocity and a reduction in the static pressure downstream from the stator. The disturbance can therefore propagate to the rotor which itself, should the conditions be sufficient, can stall as a result. The stalling also results in intense vibratory phenomena that can cause damage to the blades.

Disturbances can have multiple origins, namely from crosswind, occasional damage/asymmetry of the air inlet, stator failures from varying geometries (breakage of one or more control arms), damage due to foreign object damage (debris on runway, birds, projectiles). They can affect one or more blades; if the disturbance extends to all of the rotor blades or the stators, the affected stage can no longer function properly and carry out its compression work, while providing the necessary air mass flow for the next stage to operate. However, the pressure of the downstream stage remains higher than that of the stalled stage for at least a few moments. This results in the flow direction of the fluid being reversed, which is violently ejected towards the front of the compressor. If, after unloading the compressor, the heterogeneity of the flow has disappeared and no damage has occurred, it is possible that the compressor can be set back into operation. This usually requires the action of the pilot or of the engine control system (reduction of fuel flow, resulting in a reduction of the rotation rate). Without any corrective action, this phenomenon can be either continuous, in which case we call it a deep stall, or periodic, in the case where the compressor operates for only a short time before stalling again. In this case, we call it a compressor surge.

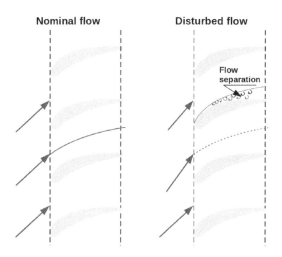

Figure 3.33. *Local effects of flow disturbance on a stator*

3.4.3.3. *The diffuser*

Located between the compressor outlet and the combustion chamber, the purpose of the diffuser is to transform some of the kinetic energy of the compressed gases into the potential energy of pressure. Indeed, the fluid velocity at the compressor outlet is too high to ensure a stable combustion in the combustion chamber (flame extinction). As the cross-sectional area of the flow gradually increases, the velocity decreases and the static pressure increases: this is the same phenomenon that was encountered for axial compressors. Consequently, it is accompanied by a total pressure loss which must be minimized. It is also important to minimize the flow heterogeneities generated by the diffuser, which may reduce the combustion efficiency.

3.4.3.4. *The combustion chamber*

As we have seen previously, the calorific value of the fuel represents the energy supplied to the heat engine. In the case of turbojet engines used in aeronautics, the common fuel is Jet A-1, a kerosene-based fuel. Its calorific value is 43.15 MJ/kg. From its combustion, the fuel provides the energy necessary to propel the aircraft.

Like any actual system, some losses occur and the useful energy available for the turbomachine is slightly lower than that in the calorific value of the fuel. Nevertheless, the gas turbine combustors (Figure 3.34) are efficient, and allow for the fuel to completely combust. Today, many combustors operate with a stoichiometric air–fuel mixture close to 1, with fuel being injected into the chamber via injectors. Air is passed through a "swirler" to create a conical shaped flame to allow for complete combustion, with fresh air being added along the walls to ensure cooling.

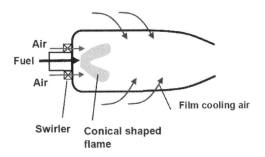

Figure 3.34. *A conventional combustion chamber*

New developments (Figure 3.35), with the aim of reducing pollutant emissions, use either "lean burn" technology, where the fuel quantity in the mixture is less than the stoichiometric proportions for combustion which, thanks to systems that allow for a good air–fuel mixture, such as a pre-mixing of fluids, allows for complete combustion. For example, "rich burn" technology, where the amount of fuel in the mixture is originally greater than the stoichiometric conditions, before being re-mixed with fresh air further downstream in the combustion chamber to burn the remaining fuel, typically known as "rich air addition" combustion ("rich burn, quick-mix, lean burn" combustion, or RQL).

Indeed, for the sake of protecting the integrity of the elements located downstream from the combustion chamber, it is not possible to leave any unburned fuel. Since the turbines are cooled by fresh air extracted from the compressor, it would be possible to cause re-ignition of the unburned fuel in the turbine and therefore subject its components to excessively high temperatures, leading to damage or their destruction.

Two types of losses can be identified in the combustion chambers used in aeronautics today.

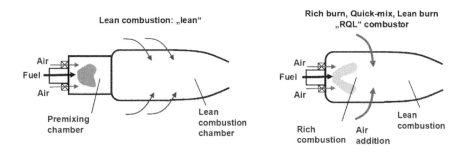

Figure 3.35. *New developments within combustion chambers*

On the thermochemical level when the fuel used is a liquid, it is necessary to vaporize it to cause its combustion. The energy required for vaporization is therefore no longer available to the gas turbine. To minimize this effect, combustion chambers are equipped with systems to inject the fuel in the form of very fine droplets, reducing the energy required for vaporization. Today, the evaluation of the energy loss due to the latent heat of evaporation is less than 1% compared to the energy available in the fuel (depending on the sources, an approximate and average value of 360 kJ/kg can be considered).

A second source of loss generated by the combustion chamber, which is not taken into account in the theoretical cycles, is the pressure losses due to friction and to the various mixing processes. Indeed, the Joule–Brayton cycle assumes that combustion occurs at a constant pressure, whereas in reality, the loss is approximately 3%.

It should also be noted that the thermodynamic combustion efficiency is not constant throughout the typical flight of an aircraft. It depends on the external conditions, in particular, altitude, pressure, temperature. This affects the conditions at the compressor outlet station (3), and therefore the conditions at the combustion chamber inlet. For example, a higher inlet pressure and temperature increases the combustion temperature, and therefore that of the hot gases (T_4) at the turbine inlet.

3.4.3.5. *The turbine*

We have seen in the Joule–Brayton cycle that an increase in the temperature at the turbine inlet increases the efficiency of the machine. In the case of a real turbine, the temperature T_4 is limited by the materials used. Even today, the quality of the materials used, such as those of monocrystalline materials or even using very sophisticated and efficient cooling systems, makes it possible to operate at very high temperatures, up to approximately 1,800—1,900 K for cooled monocrystalline materials. However, this temperature limit T_4 cannot be increased indefinitely.

On the other hand, in the Joule– Brayton cycle, the expansion phase that takes place in the turbine is considered to be adiabatic and reversible, i.e. isentropic.

This assumes that there are no losses generated in the turbine. In reality, various loss-inducing phenomena are observed.

Since turbojet turbines are not isolated from the exterior (Figure 3.36), their operation is therefore not perfectly adiabatic. The air circulating between the nacelle and the casing is cold, which also keeps the casing at a temperature low enough to maintain high mechanical characteristics, resulting in heat being lost to the outside. This energy that is lost through the turbine cannot therefore be transformed into mechanical work.

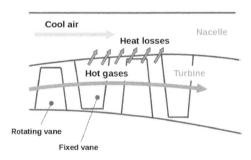

Figure 3.36. *Real non-adiabatic turbine*

A second source of loss is linked to the frictional effects between the gas and the walls of the turbine, blades and platforms. Indeed, since air is a viscous fluid, the friction generated is therefore non-zero, contrary to the isentropic approximation for the expansion. Any friction generated tends to reduce the kinetic energy of the fluid, energy which therefore cannot be transformed into mechanical work.

The additional losses that should be mentioned are related to the actual appearances of such machines. Indeed, between each rotating and fixed part, a whole system of clearances allows for the machine to operate in any condition, without creating mechanical friction and avoiding contact between the parts. These clearances cause leaks, which often allow the gases to follow a path without providing work. This phenomenon appears, for example, in the case of a rotating blade. This blade rotates in a fixed ring called the casing and, to prevent the blade from making contact with the casing and thus deteriorating, a certain amount of clearance is left available. This causes an airflow to form between the blade and the casing, which does not contribute to the work done by the machine. Of course, many design evolutions are used to reduce this leakage rate. For a high-pressure turbine blade subject to very high stresses, the blade tip geometries have a three-dimensional contour allowing for air recirculation to form, and thus reduce leakages (Figure 3.37).

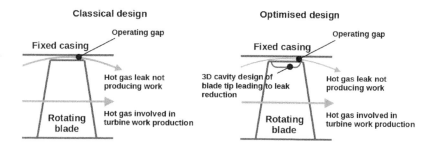

Figure 3.37. *Clearances between rotating blades and casing in a turbine*

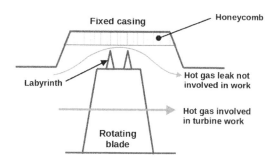

Figure 3.38. *Shrouded rotating blades*

In low-pressure turbines, designs have evolved towards using shrouded blades (Figure 3.38). Here, the blade edges are decorated with teeth facing a part of the casing having had a honeycomb added to it. This system considerably reduces the leakage flow, while also reducing its impact on the main airflow. Indeed, the shrouded-style geometries make it possible to guide the leaks more easily, and thus avoid the creation of recirculation flows from the vortex style of the blade tip. These vortices are created by the passage of fluid between the pressurized intrados and the depressurized extrados.

In order to maintain the turbine at a temperature that allows it to operate and last long enough for industrial use, we have also mentioned that an ingenious cooling system is used. The secondary air system (Figure 3.39) derives air from colder stations, and injects it into the turbine to ensure its cooling, or to pressurize certain cavities in order to avoid any hot gas suction. This secondary air system mixes locally fresh air with the hot gases, which also induces losses from the mixing phenomena. Indeed, these gases have different kinetic energies, rotational speeds and temperatures.

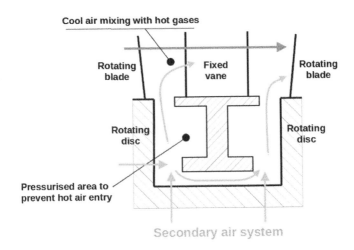

Figure 3.39. *Cavity between a disk and a fixed vane in a turbine*

3.4.3.6. *The exhaust nozzle*

The exhaust nozzle, located downstream of the turbine, constitutes the last element of the engine, provided that we disregard the afterburner systems, thrust reversers or installations related to noise attenuation/suppression (mixers, chevrons).

Theoretically, the gases leaving the turbine could be ejected directly into the atmosphere. The addition of a convergent-shaped exhaust nozzle makes it possible to increase the velocities of the gases at the outlet of the engine, and therefore increase the propulsive efficiency by increasing the difference in the velocities between the air entering the reactor and the air leaving the reactor. What is more, the velocities of the gases leaving the turbine have a strong circumferential component relative to their axial component: the nozzle thus acts as a rectifier.

The gas ejection nozzle is therefore generally a circular or oval pipe, with a decreasing cross-sectional area containing shrouded stators. The maximum axial speed is obtained at the level of the minimum cross-sectional area, where the static pressure is also at its minimum (ejector effect). The minimum cross-sectional area, or throat, is designed so that the gas velocity remains subsonic, and that during the engine operation during the cruising phase, the static pressure at the air jet outlet is the same as the atmospheric pressure: all of the potential energy of the gas will have been converted into kinetic energy.

Obtaining a sonic velocity at the throat is not desirable for a convergent nozzle, as this would result in too much of an efficiency loss.

In some cases (military applications), the nozzle is convergent–divergent (*Laval nozzle*). In this case, a sonic velocity is actively sought after at the throat of the nozzle, and the nozzle is said to be primed. Via supersonic expansion, the divergent nozzle will further increase the velocity of the gases beyond the speed of sound (Hugoniot's conditions).

Some nozzles have a variable geometry whose outlet cross-sectional area changes according to the fuel flow, among other things, thus adapting for each phase of flight.

3.4.4. *Secondary air system design methods*

The *secondary air system* is a complex system for routing air from specific stations within the engine, usually in the compressor region, to the hot areas of the turbine. The development of such a system is carried out in different stages from the overall design, with a detailed analysis of certain critical positions.

In the ideal case, the air is extracted from the compressor at the desired pressure and routed to the turbine to provide the desired use: mainly cooling or pressurizing. The pressure level of the secondary air system is decisive, because it is chosen in such a way as to ensure the necessary tightness, without overestimating the expected air pressure and therefore avoiding greater losses. It is important to remember that approximately 20% of the total turbojet flow is used in the secondary air system, and therefore does not participate in the engine cycle.

In the real case, air is transported from various parts of the engine, through different orifices, pipes, between rotating components, etc. All of these obstructions generate pressure losses, which should be taken into account when designing the complete system.

Each stage of the development cycle of the secondary air system is summarized in Figure 3.40.

For the design of the secondary air system, different specifications and fresh air requirements are evaluated. A zero-dimensional (0D) system is set up to determine which sources are suitable for which needs. From this, a one-dimensional (1D) system is then built in such a way as to simulate the routing of the fluid, simulating the obstacles using characteristics of pressure drops according to the operating points of the engine.

Once all of the pressure drops have been defined and correspond to the real topology of the engine, it is integrated into a 2D thermal model which is undertaken in order to refine the predictions by taking the temperatures into account. This thermal model, in addition to accounting for the fluid mixtures at different temperatures, makes it possible to simulate the heat transfer within the materials of the components, simulating the energy exchange with the surrounding fluid.

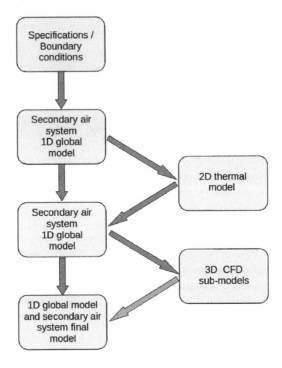

Figure 3.40. *Secondary air system development cycle*

The final step is to increase the precision of the numerical models, mainly in areas where the uncertainties are non-negligible, and the effects are direct. This last step is the simulation of sub-models, i.e. specific parts using 3D models. These calculations are generally carried out using computational fluid dynamics (CFD) simulations based on the Navier–Stokes equations. These calculations make it possible to increase the precision of the estimates, compared to the 1D and 2D thermal models, by making corrections to the empirical pressure drops that were previously used. Indeed, in general, 1D and 2D thermodynamic simulation software is mainly based on correlations and characteristics from measurement campaigns, carried out most of the time using more theoretical geometries.

3.4.5. T_4 *and the secondary air system*

As discussed in section 3.3, to increase the efficiency of the Joule–Brayton cycle, it is advisable to increase the area between the isentropes and the isobars by as much as possible. A direct way to achieve this result, at least theoretically, is to artificially increase the temperature T_4, at the turbine inlet.

An increase of T_4 is accompanied by a temperature increase of the materials (blades) directly placed in the hot airflow. This increase can be initially done by using materials which are resistant to high temperatures. This was widely practiced from the 1960s when the development of new materials, in conjunction with new casting techniques, achieved considerable success.

However, the development of increasingly high-performance materials is a costly and time-consuming activity, which involves many risks during the industrialization phase (moving from the prototype to large-scale use in a series). Improvement of the cycle must therefore be very rapid since customers (aircraft manufacturers and airlines) are mostly interested in the more economical and ecological engines.

The required turbine inlet temperatures soon exceeded the maximum permissible temperatures that the materials could attain, and thus it was necessary to turn to alternative technologies. Today, the maximum air temperature at the turbine inlet can exceed 1,600°C.

The initial idea was to cover the blades with a coating that would act as a thermal insulator, and thus reduce the heat exchange between the hot airflow coming from the combustion chamber and the blades, by as much as possible. The second idea was to cool the blades continuously, in order to guarantee an acceptable maximum temperature of the materials.

Cooling is provided by secondary air systems. In an aeronautical turbojet, there are two air systems.

The main (or primary) air system is made up of all the airflows contributing to the generation of thrust. For dual-flow architectures, there are two possible forms:

– The primary hot flow which passes through each of the core modules of the engine (compressor, combustion chamber, turbine), and on which most of the work done by the turbomachine is carried out, but only contributes to about 20% of the thrust.

– The secondary cold flow which, once accelerated by the fan, avoids the core modules of the engine and is ejected at the rear end, which is responsible for 80% of the thrust.

The secondary air system brings together all of the airflows which are not used to generate thrust. However, its role should not be overlooked. The secondary air system is itself subdivided into two categories:

– The external secondary system accounts for any flow of air that does not directly serve the engine. This system is responsible for:

    – the pressurization/air conditioning (fixed pressure altitude from an altitude of 1,850 m for the A380) of the cabin/hold. Note that, for passenger comfort, pressurization must recreate a low/medium altitude environment, whereas from a structural point of view, it would be ideal not to pressurize the cabin (no pressure force on the structure),

    – for de-icing the leading edges of the wings and engine air intakes.

– The internal secondary air system covers any airflow involved in the proper functioning of the turbojet engine. This includes:

    – the active cooling processes of the turbine blades, the external casing of the compressor and the recently discussed turbine,

    – the passive cooling processes of the disks, on which the blades are installed,

    – the pressurization processes for the inter-stage cavities, to ensure that no ingestion of hot gas from the primary flow occurs which would increase the disk temperature,

    – the pressurization processes for the bearing chambers, so as to prevent any spillage of the lubricating liquid onto the hot parts (risk of oil fires),

    – the pressurization processes of the modules, in order to prevent excessive axial forces at the rolling bearings and increase their life span.

The secondary air, taken directly from the primary flow, must be compressed to fulfill its various tasks. It is sampled from different points along the compressor, with each air sample being extracted at a pressure level chosen to most properly suit its end use. For cabin pressurization, the air will be taken from the first stages of the low-pressure compressor, while for the cooling of the blades located at the inlet of the turbine ($p_4 = p_3 - \Delta p_{combustion}$), air will be taken from the final stages of the compressor.

The "cost" of producing secondary air is therefore very different depending on its different uses. In any case, it is advisable to minimize the quantity of air diverted to be used in the secondary air system, which can affect the quality of the entire cycle in a significant way.

Contrary to the simplifying assumptions of the previous parts, the airflow W2 at the compressor inlet is not maintained until station 3, but occasionally decreases throughout its course in the compressor, and occasionally increases during its journey through the turbine.

One of the functions of the secondary air system is to provide gap active clearance system, which indirectly improves the thermodynamic efficiency of the engine system.

So far, we have assumed that all of the air entering through station 2 (compressor inlet), which constituted the main airflow, was used either to generate the rotation of the compressor by means of the turbine, or to generate an increase in the airflow velocity at the nozzle outlet.

By introducing the concept of a secondary air system, we started to challenge this assumption, and we will continue to do so. If part of the air entering into the compressor does not reach the inlet of the combustion chamber, the work which has been done on this air mass does not contribute to either the driving of the turbine, or the increase of the velocity at the outlet of the engine: this constitutes a loss of thermodynamic efficiency.

The same applies to any air mass passing through the compressor without being compressed, or through the turbine without any contribution to driving the blades.

Consider a cross-section of the compressor or turbine for one rotor stage (Figure 3.41). The internal engine housing forms a sealed envelope through which air cannot escape. The engine casing is fixed, and the vanes, connected to the disk, are rotating. So there is a gap between the inner housing of the casing and the outer diameter of the vanes, between which any contact would cause unacceptable damage.

Even if the spacing was very small (< 1 mm), at such a large diameter the resulting area is significant: some of the air will be able to pass between the casing and the vanes, and will therefore not contribute towards the thermodynamic cycle. It is necessary therefore to minimize these gaps, which give rise to parasitic flows, by as much as possible.

To complicate things a little more, this gap, which is inherent in the construction of the engine, does not remain constant during the different phases of flight.

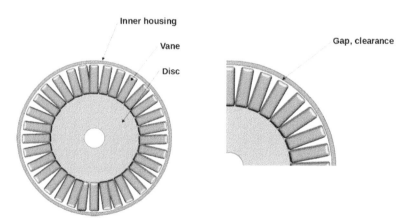

Figure 3.41. *Cross-section of the engine, at the level of a rotor stage*

Indeed, if we consider the take-off phase, the following events can be observed within the turbine:

1) The pilot sets the throttle to maximum power.

2) Via the on-board computer, the fuel–air mixture will be enriched with fuel, which increases the temperature experienced by the vanes and blades. Under the effect of the almost instantaneous change in temperature, the vanes and blades or airfoils, submerged in the hot airflow, will initially expand, reducing the spacing between them and the casing.

3) As the energy level of the gases entering the turbine increases, the vanes will be able to recover more mechanical energy. This will increase the rotational frequency and therefore the centrifugal forces on the vanes, which will then expand. The spacing is once again reduced.

4) The casing starts to feel the effects of the temperature change. It can be considered as a thin ring which tends to increase in diameter due to the effect of thermal expansion: the spacing between the vanes and the casing thus increases.

5) If this phase continues for long enough, the disk itself will undergo a temperature increase and will also expand, along with the vanes, by thermal conduction. This will once again reduce the spacing.

Since the take-off phase is short in general, the phenomena described in point 5 does not really appear. Therefore, we will not attempt to compensate for these gaps during this flight phase, in addition to any other similar ones. If we consider another flight phase, for example, cruising, which is characterized by an almost constant engine velocity over a long period of time, the control of the gaps becomes more important. Indeed, the benefits of reducing the gaps between the vane tips and the casing will be substantial in reducing the specific fuel consumption.

So, how can we control the gaps? The most widespread system consists of regulating the thermal expansion of the casing by cooling it down from its exterior. To do this (Figure 3.42), channels are placed around the casing with a small spacing (of the order of a few millimeters). These channels are pierced on their interior diameter to form several orifices, through which fresh air can be blown onto the casing in order to cool it by forced convection. These channels are usually placed either directly above or in a close proximity to the rotating vanes, as to achieve a directly controllable tightening effect. The gap active control system is installed in the compartment between the engine and the bypass channel. The prevailing pressure in this region is sufficiently close to the local atmospheric pressure at a cruising altitude. Therefore, we can take the cooled air from the first stages of the compressor, thereby limiting the losses coming from the bypass of compressed air.

The problem of the gaps is found both in the compressor and in the turbine. For double- or triple-spool engines, it is easy to imagine a specific clearance control system for each module (compressor/low- and high-pressure turbine). However, it

Aeronautical and Space Propulsion   131

is necessary to analyze the benefits of using such a system critically, as well as the negative effects that it entails:

– the logic for opening and closing the air control valve, in order to operate the gap active control system in a specified manner;

– an increase in the complexity of the system, and therefore potentially the frequency of failures;

– the installation and size of the spacing between the engine and the bypass channel;

– the additional mass;

– the air bypass.

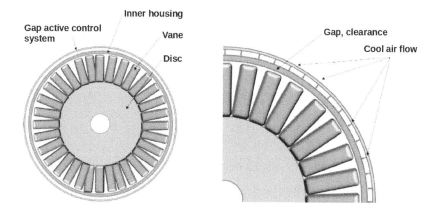

Figure 3.42. *Schematic view of gap active control system (front view)*

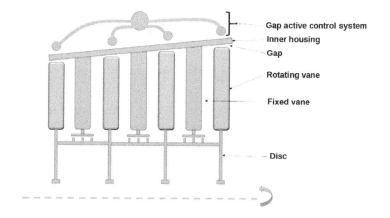

Figure 3.43. *Schematic view of a low-pressure turbine (profile view)*

3.5. Perspectives

Mass reduction, higher temperature-resistant materials and limiting the environmental impact (consumption, pollution, noise) of engines are all major challenges in aeronautics.

Increasingly elaborate programs based on the numerical modeling of flows and heat transfers in the various components of aeronautical engines have been developed for many years by manufacturers and public research organizations (DLR in Germany, ONERA in France, etc.).

The "Clean Sky Joint Undertaking" project has more than 600 partners, and is the largest European aeronautical research project currently aimed at developing innovative techniques to reduce the emissions of CO_2, and other gases coming from aircraft, by developing a set of technologies necessary for "a clean, innovative and competitive air system".

3.6. References

Abbott, I.H. and von Doenhoff, A.E. (1959). *Theory of Wing Sections*. Dover Publications.

Bauer, P. (2009). *Propulseurs aéronautiques et spatiaux*. Technosup, Ellipses.

Bouchez, M. (2018). Propulsion aérospatiale – Introduction. *Techniques de l'Ingénieur*, BM 3 000v2, 1–13.

Farokhi, S. (2014). *Aircraft Propulsion*, 2nd edition. Wiley.

Giovannini, A. and Airiau, C. (2016). *Aérodynamique fondamentale*. Cépaduès.

Hill, P. and Peterson, C. (1992). *Mechanics and Thermodynamics of Propulsion*, 2nd edition. Addison-Wesley.

Milne-Thomson, L.M. (1958). *Theoretical Aerodynamics*, 4th edition. Dover Publications.

SAE Aerospace (2013). Aircraft Propulsion System Performance Designation and Nomenclature. SAE Aerospace Recommended Practice, SAE ARP755C.

Sforza, P.M. (2016). *Theory of Aerospace Propulsion*, 2nd edition. Butterworth Heinemann.

4

Combustion and Conversion of Energy

Bernard DESMET

INSA – HdF, Université Polytechnique Hauts-de-France, Valenciennes, France

4.1. Generalities

4.1.1. *Introduction*

Temperatures greater than that for the atmosphere are obtained in the majority of cases by *combustion*, i.e. by an *exothermic chemical reaction* between a *reducing agent* and *oxygen* (or another *oxidizing agent*).

Combustion should be distinguished clearly from *pyrolysis* (or thermolysis), which is the chemical decomposition of an organic compound in the absence of oxygen, or in a poor atmosphere, to differentiate between oxidation and combustion. Pyrolysis, which is a way to synthesize various solid, liquid or gaseous hydrocarbons, can be used to transform biomass into energy sources that are easier to use. Pyrolysis begins at a relatively low-temperature level ($\approx 200°C$) and continues up to about $1,000°C$.

Combustion can only occur if the following three conditions are satisfied (Figure 4.1):

– A *fuel*, which can be in either solid, liquid or gaseous form, contains the chemical species to be oxidized: carbon (C), hydrogen (H), sulfur (S), etc., and sometimes also oxygen (O) and nitrogen (N).

Thermodynamics of Heat Engines,
coordinated by Bernard DESMET.
© ISTE Ltd 2022.

– The *oxidizing agent* or *oxidizer*, which contains the oxygen required for combustion. Most of the time, the oxidizer is the air, or more particularly, the oxygen constituted within the air. The products of the combustion reaction constitute the *combustion gases*, also sometimes called "smoke" or burnt gases.

– An *activation energy* (heat or a spark, for example), which is the trigger to initiate the combustion reaction.

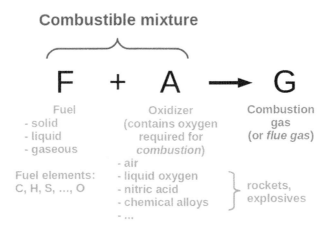

Figure 4.1. *A combustion reaction*

Slow combustion, which produces only a weak release of heat, differs from *rapid combustion*, and can result in either:

– a *flame*: when the oxidation reaction within a small thickness produces heat and, in general, with an emission of light;

– an *explosion*: *deflagration* or *detonation*, if the speed of the flame front exceeds that of sound.

In the remainder of this chapter, the term "combustion" will be reserved for live combustion taking place and producing a flame.

The thermal energy released by combustion can be used in many applications:

– Production of heat:

    - boilers: heating, steam production (thermal power plant cycles, food industry, etc.),

    - ovens: cement and brick works, heat treatment, etc.,

    - chemical processes,

    - etc.

– Production of mechanical energy cycles:
  - gas turbines,
  - propulsion cycles (turbojets, rocket engines),
  - internal combustion engines,
  - etc.

Most often, combustion is carried out at constant pressure on a continuous basis. In the case of internal combustion engines, the combustion process is much more complex, because it is carried out under conditions with varying pressures and volumes in a non-steady state.

4.1.2. Premixed flame

The situation represented in Figure 4.2 corresponds to the propagation of a flame in a *homogeneous mixture* made up of a fuel, for example, methane (CH_4) and air.

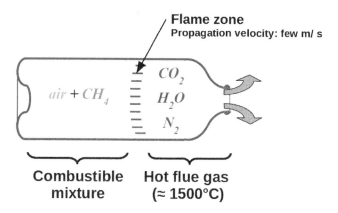

Figure 4.2. *Premixed flame*

The flame, initiated at the level of the neck and which can be observed from the light emitted, progresses in the combustible mixture at a relatively low speed of a few m/s. The hot gases produced by the combustion, which have a temperature of the order of 1,500°C, are ejected outside.

This is the type of combustion that occurs in spark-ignition engines (see sections 2.1.2 and 2.4.6), in which the air–fuel mixture contained in the cylinder is formed before the start of combustion, and whose activation energy is generated by an electric spark from the spark plug. This type of combustion will be studied in more detail later.

In the case of a Bunsen burner (Figure 4.3), combustion is carried out continuously. The fuel and the air introduced circulate in the tube for long enough, to allow the homogenization of the combustible mixture. The flame is then stabilized at the end of the mixing tube.

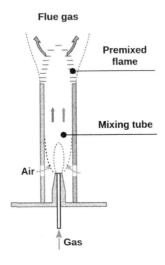

Figure 4.3. *Premixed flame from a Bunsen burner*

4.1.3. Diffusion flame

In the case of a diffusion flame, for example, the flame of a gas lighter (Figure 4.4), the fuel and oxidizer are mixed together within the flame zone, with the flame being located in the zone where the proportions within the mixture allow for oxidation to occur. Temperatures in the flame area can be very high.

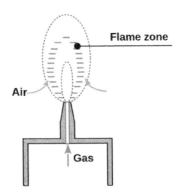

Figure 4.4. *Flame from a gas lighter*

In the case of a diesel engine (see section 2.1.2), a diffusion flame develops in the mixing zone for the jet of fuel, injected at a high pressure, and with the air contained in the cylinder. The high temperature of the air, which has been strongly compressed, provides the required activation energy for the fuel to self-ignite.

Diffusion flames can be produced in a steady state, as in the case of a combustion chamber (Figure 4.5), and can be found in the mixing zone of the fuel along with the primary air. Hot gases with high temperatures are mixed with the dilution air, in order to cool down the gases at the combustion chamber outlet to a level compatible with the equipment located downstream from the combustion chamber. For example, the turbine in the case of a gas turbine installation.

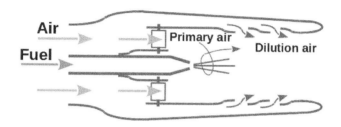

Figure 4.5. *Diffusion flame in a combustion chamber*

4.1.4. *Stabilization of a flame*

The presence of a suitable shaped body (in the case of Figure 4.6, a torus) in the flow of the combustible mixture produces a *recirculation vortex* of the gases at high temperatures, serving as an initiator for the oxidation reaction. The flame is then anchored by this recirculation vortex (Borghi and Destriau 1995).

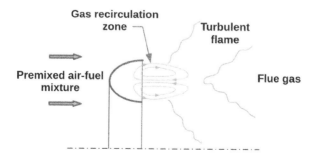

Figure 4.6. *Stabilization of a flame by anchoring*

4.1.5. *Flammability of air–fuel mixtures*

To be flammable, the fuel must be present in the form of a gas or a vapor. The flammability of a fuel is determined by its *ability to ignite and sustain its combustion*. The range of fuel–air concentrations in which combustion is possible is limited; for concentrations below and above that, for which ignition and flame propagation are not possible (Figure 4.7), are, respectively, the lower (LFL) and upper (UFL) *flammability limits*. In practice, these flammability limits are generally expressed for air mixtures as a function of the pressure and temperature. For most hydrocarbons, the lower limit decreases with increasing temperature, while the upper limit increases.

In the absence of an ignition source, a combustible mixture within the flammability limits may ignite spontaneously just from the temperature, which is defined as *minimum auto-ignition temperature* (T_{MI}).

Flammability is also characterized by the temperature of the *flash point* T_F, i.e. the minimum temperature for which a fuel evaporates and then mixes with air, igniting on the surface of the liquid fuel.

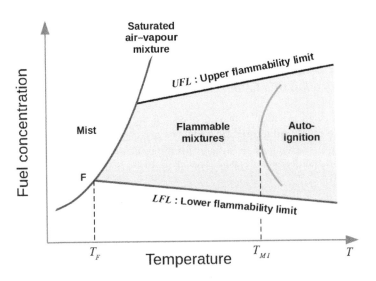

Figure 4.7. *Flammability range as a function of temperature, according to Zabetakis (1965)*

Table 4.1 gives the flammability limits of some air–fuel mixtures at the standard atmospheric pressure and ambient temperature (20°C), expressed in percentages by volume as well as the minimum auto-ignition temperatures, and temperature of the flash point.

	Lower inflammability limit $LFL\ [\%]$	Upper inflammability limit $UFL\ [\%]$	Minimum auto-inflammation temperature $T_{MI}\ [°C]$	Flash point temperature $T_F\ [°C]$
Hydrogen	4	75	585	
Methane	5	15	537	-188
Propane	2,4	9,3	450	-104
n-Octane	0.95	3.20	220	13
Iso-octane	0.79	5.94	417	4.5
Essence (octane ind. 100)	1.4	7.6	246	< - 40
Diesel	0.6	7.5	210	> 62

Table 4.1. *Flammability properties of some air–fuel mixtures*

4.1.6. *Combustion in internal combustion engines*

It is important to distinguish between the case of a spark-ignition engine from that of a compression ignition engine (Griffiths and Barnard 1995).

In the case of *spark-ignition engines* – in particular gasoline engines – the combustible mixture is formed during the intake phase or at the start of compression. Combustion is then initiated by an electric spark when it should occur, namely when the piston is near the top dead center (Figure 4.8).

A premixed flame then spreads through the fresh gases from the point of ignition. After compression, the temperature of the gases must remain below the auto-ignition temperature for the combustible mixture, to prevent any premature combustion from occurring prior to the spark from the spark plug, which explains why the compression volume ratio must be limited for controlled ignition engines. In practice, the compression volume ratio for automotive gasoline engines is of the order of 10. The *knocking phenomenon* degrades the combustion process which can occur, in particular, if the compression ratio is too large or if the temperature is too high: combustion begins as usual but the pressure and temperature increase causes the self-ignition of the fuel mixture in certain regions, in particular near the walls, causing part of the fuel mixture to ignite prior to the flame front reaching this region and is accompanied by vibrations in the frequency range of the order of 5–10 kHz. This phenomenon is likely to cause damage to the piston, segments and the cylinder, and so should be avoided. The octane number characterizes the resistance that a fuel has to induce this knocking phenomenon.

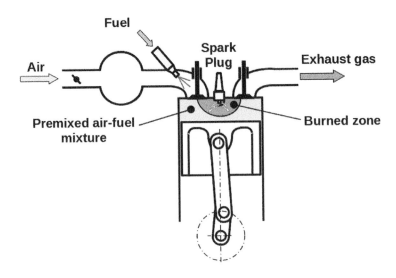

Figure 4.8. *Combustion in a spark-ignition engine*

In the case of *compression ignition engines* (Figure 4.9), air is introduced into the cylinder at full intake, and is isolated from any fuel during the compression phase. Fuel is then injected into the cylinder at a high pressure when the combustion should occur, namely, the moment when the piston is near top dead center. When fuel is injected into the cylinder, the gases inside must be at a high enough temperature to cause the fuel to self-ignite, which imposes the condition on the volumetric compression ratio that is has to be of the order of 20 in the case of an automotive diesel engine. A diffusion flame develops in the outer zone of the jet where the air is sufficient in quantity to ensure that the fuel will oxidize. The local temperatures in the combustion zone are very high, and play an important role in the formation of polluting species (*nitrogen oxides*).

Engines operating in the Homogeneous Charge Compression Ignition (HCCI) mode of combustion have been the subject of research for several decades. The filling of the air–fuel mixture is carried out at full intake, before then being compressed with a sufficient volumetric ratio that meets the self-ignition conditions. Combustion then occurs from multiple sites situated throughout the combustion chamber, although without a very high-temperature zone as in the case of the diffusion flame in a diesel engine. At low temperatures, HCCI combustion essentially does not produce any nitrogen oxides (NOx), nor does it generate many particles. On the contrary, the knocking phenomenon occurs when the fuel proportion in the combustible mixture exceeds a certain limit, thus restricting this combustion mode solely to operations with medium loads and, to operate at higher loads, it is necessary then to return to conventional means of operation.

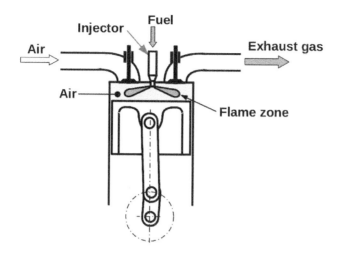

Figure 4.9. *Combustion in a compression ignition engine*

The study of combustion within engines is very complex, since:
– fuel constitutes a mixture of several chemical species;
– the combustion process involves many intermediate reactions;
– combustion occurs in a variable volume, with heat exchanges (cooling) and work being done in an unsteady state.

4.2. Theoretical combustion reactions

4.2.1. *Constituents of the combustible mixture*

4.2.1.1. *Fuel*

Fuels are generally made up of a mixture of many chemical species containing carbon and hydrogen, such as alkanes (C_nH_{2n+2}) and possibly with other species containing oxygen, such as ethanol (C_2H_5OH) or nitrogen such as hydrazine (N_2H_4).

For fuel that contains carbon, hydrogen and oxygen, a chemical representation in the following form is possible:

$$\text{Fuel: } C_xH_yO_z \quad\quad [4.1]$$

where the coefficients x, y, z, which may not necessarily be integers, can be defined in such a way as to respect the proportions of the chemical species contained in the fuel mixture, without respecting the average molecular mass of its constituents.

4.2.1.2. Combustion air

In most cases, combustion is carried out using the air from the atmosphere as the oxidizer. The composition of air by volume is principally nitrogen (N_2 : 78.08%) and oxygen (O_2 : 20.95%), as well as rare gases constituting less than 1% of the mixture (argon: 0.93%, neon: 0.0018%, krypton, xenon, helium).

By replacing the rare gases, whose influence on combustion can be neglected, with an equivalent quantity of nitrogen and, according to the ideal gas equation of state, the volume and molar proportions are equal, allowing us to represent the air composition by the chemical formula:

$$\text{Air: } O_2 + 3.76\, N_2 \qquad [4.2]$$

In the lower layers of the atmosphere, the air also contains water vapor and carbon dioxide (CO_2).

4.2.2. Combustion stoichiometry

Combustion is said to be *stoichiometric*, or neutral, when the fuel mixture contains the exact quantity of oxidizer needed for the complete oxidation of the fuel, i.e. to produce only carbon dioxide CO_2 and water H_2O.

The theoretical stoichiometric combustion of fuel ($C_x H_y O_z$) using air ($O_2 + 3.76\, N_2$) as oxidizer corresponds to the case where the there is just enough oxygen to allow for the fuel to completely oxidize. Gases arising from the combustion are composed of carbon dioxide (CO_2), water (H_2O) and nitrogen (N_2). As nitrogen is an inert gas however, it does not take part in the chemical reaction for combustion. Under these conditions, the overall combustion reaction is of the form:

$$C_x H_y O_z + n\,(O_2 + 3.76\, N_2) \longrightarrow a\, CO_2 + b\, H_2O + d\, N_2 \qquad [4.3]$$

where the coefficients n, a, b, d are obtained easily, by balancing the reaction:

$$C_x H_y O_z + \left(x + \frac{y}{4} - \frac{z}{2}\right)(O_2 + 3.76\, N_2) \longrightarrow$$
$$x\, CO_2 + \frac{y}{2} H_2O + 3.76\left(x + \frac{y}{4} - \frac{z}{2}\right) N_2 \qquad [4.4]$$

These coefficients correspond to the stoichiometric proportions.

Given these coefficients for the combustion equation, the ratio A/F of the mass of air to the fuel mass can be defined. In the case of stoichiometric combustion, this ratio is:

$$\left(\frac{A}{F}\right)_{th} = \frac{\left(x + \frac{y}{4} - \frac{z}{2}\right)(M_{O_2} + 3.76\, M_{N_2})}{x\, M_C + y\, M_H + z\, M_O} \qquad [4.5]$$

The *molar masses* M_c, M_H, M_O, M_N used in equation [4.5] are provided in Table 4.2.

Molar masses [g/mol]			
Carbon : C	Hydrogen : H	Oxygen : O	Nitrogen : N
M_C : 12.011	M_H : 1.0078	M_O : 15.999	M_N : 14.007

Table 4.2. *Molar masses*

To characterize the proportions of a real combustible mixture, we generally define its *air factor* λ, or its *richness* R, which is the inverse of the air factor:

$$\text{Air factor: } \lambda = \frac{\dfrac{A}{F}}{\left(\dfrac{A}{F}\right)_{th}} \qquad \text{Richness: } R = \frac{1}{\lambda} = \frac{\dfrac{F}{A}}{\left(\dfrac{F}{A}\right)_{th}} \qquad [4.6]$$

The excess air e can also be defined:

$$\lambda = 1 + e \qquad [4.7]$$

An air factor $\lambda > 1$ ($R < 1$: lean mixture) therefore corresponds to combustion occurring with an excess of air, whereas if $\lambda < 1$ ($R > 1$: rich mixture), then combustion occurs with an excess of fuel.

The richness of a fuel mixture depends on the technologies used:
– $R \approx 0.98$ to 1.02: petrol engines;
– $R \approx 0.10$ to 0.80: diesel engine and turbojets;
– $R \approx 0.50$: HCCI engines.

4.2.3. *Theoretical combustion of a lean mixture*

In the case when a lean mixture ($\lambda > 1$; $R < 1$) combusts, the quantity of oxygen available is greater than which is needed to ensure that the fuel completely oxidizes (oxidative combustion). The excess oxygen then remains contained within the combustion gases. The theoretical reaction for this combustion can be written in the form:

$$C_xH_yO_z + \lambda \left(x + \frac{y}{4} - \frac{z}{2}\right)(O_2 + 3.76\, N_2) \longrightarrow$$
$$x\, CO_2 + \frac{y}{2} H_2O + (\lambda - 1)\left(x + \frac{y}{4} - \frac{z}{2}\right) O_2 \quad \text{[4.8]}$$
$$+ 3.76\, \lambda \left(x + \frac{y}{4} - \frac{z}{2}\right) N_2$$

Equation [4.8] shows that oxygen is the only chemical species which reacts with the fuel, according to the chemical equation:

$$C_xH_yO_z + \left(x + \frac{y}{4} - \frac{z}{2}\right) O_2 \longrightarrow x\, CO_2 + \frac{y}{2} H_2O \quad \text{[4.9]}$$

and the excess gases contained within the air:

$$\left(x + \frac{y}{4} - \frac{z}{2}\right)\left((\lambda - 1) O_2 + 3.76\, \lambda\, N_2\right) \quad \text{[4.10]}$$

do not undergo any chemical reaction during the theoretical combustion.

4.2.4. *Theoretical combustion of a rich mixture*

In the case when a rich mixture ($\lambda < 1$; $R > 1$) combusts, the quantity of oxygen is insufficient to ensure the complete oxidation of the fuel (reductive combustion). The carbon contained within the fuel cannot be completely oxidized, and is found partly in the form of either carbon dioxide (CO_2) or carbon monoxide (CO) in the burned gases. The theoretical reaction of combustion can be written in the form:

$$C_xH_yO_z + \lambda \left(x + \frac{y}{4} - \frac{z}{2}\right)(O_2 + 3.76\, N_2) \longrightarrow$$
$$\left(x - 2(1-\lambda)\left(x + \frac{y}{4} - \frac{z}{2}\right)\right) CO_2 + 2(1-\lambda)$$
$$\left(x + \frac{y}{4} - \frac{z}{2}\right) CO + \frac{y}{2} H_2O + 3.76\, \lambda \left(x + \frac{y}{4} - \frac{z}{2}\right) N_2$$
$$\text{[4.11]}$$

4.3. Energy study of combustion

4.3.1. *Combustion at constant volume*

Here, we will consider the case of a mass m of fuel combusting *in the presence of excess air* (lean mixture – $\lambda > 1$) initially at the pressure p_0. The initial conditions are the quantities of fuel and air present, which are contained in a non-deformable tank, assumed to be at a temperature T_0 chosen as reference (Figure 4.10). In general, the reference temperature is taken to be $T_0 = 25°C$.

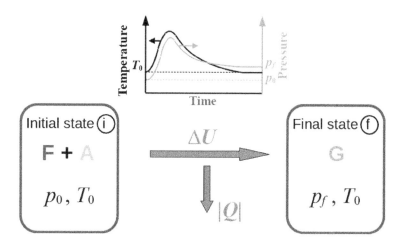

Figure 4.10. *Combustion at constant volume*

Once the amount of heat $|Q|$ has been extracted from the considered system, the temperature of the combustion products is reduced to the value T_0 in the final state. Note that the final pressure p_f of the combustion gases may be different from the initial pressure p_0, which, in particular, results from the fact that the number of moles varies due to the chemical oxidation reaction (phenomenon of "molecular expansion"). ΔU represents the variation of internal energy of the system between the initial and final states.

As the combustion occurs in a non-deformable tank, no work is exchanged with the exterior and, according to the first law of thermodynamics (equation [1.2]):

$$Q = \Delta U \quad (Q < 0) \qquad [4.12]$$

As noted in section 4.2.3, the excess oxygen and nitrogen contained in the air do not undergo any chemical reaction during combustion. By considering these gases as ideal or semi-ideal gases, their internal energy depends solely on the temperature, and therefore, its variation during the evolution in question is zero. Therefore, the variation in internal energy ΔU results only from those that have undergone a chemical reaction (equation [4.9]), depending only on the mass m of the fuel present. The heating value at a constant volume can then be defined by:

$$HV_v = \frac{|\Delta U|}{m} \qquad [4.13]$$

Depending on the state of the water formed from combustion, a distinction must be made between:

– the *lower heating value at a constant volume* LHV_v, if the water formed is considered to be in its vapor phase at the reference temperature T_0;

– *the higher heating value at a constant volume* HHV_v, if the water formed is considered in its liquid phase at the reference temperature.

The difference between the upper and lower heating values corresponds to the *latent heat of vaporization* at the temperature T_0 of the water formed by the combustion ($l_v = 2,441.7$ kJ/kg, for the temperature $T_0 = 25°C$).

4.3.2. *Combustion at constant pressure*

We assume that the conditions are identical to those indicated in the previous section, except for the volume of the tank which we can modify by moving a piston, so as to ensure a constant pressure p_0 during the combustion (Figure 4.11). According to the first law of thermodynamics (equation [1.2]):

$$\int_{(i \to f)} -p_0 \, dV + Q = \Delta U \quad (Q < 0) \qquad [4.14]$$

which can also be written in the form:

$$Q = (U_f + p_f V_f) - (U_i + p_i V_i) = \Delta H \quad \text{with: } p_f = p_i = p_0 \qquad [4.15]$$

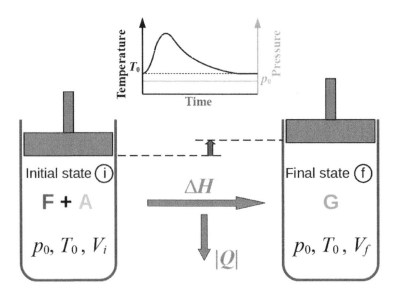

Figure 4.11. *Combustion at a constant pressure*

where ΔH represents the variation in enthalpy for the species which undergoes the chemical reaction during the combustion and, analogously to what was stated in the case of combustion at a constant volume, the heating value at a constant pressure can be defined by:

$$HV_p = \frac{|\Delta H|}{m} \qquad [4.16]$$

and, depending on the whether the water formed is in its liquid or vapor state, the *higher heating value at a constant pressure* HHV_p should be distinguished from the *lower heating value at a constant pressure* LHV_p.

4.3.3. *Relations between heating values*

4.3.3.1. *Heating value at a constant pressure*

The heating values of fuels at a constant pressure can be derived from the *standard enthalpies of formation* for the chemical species involved in the combustion reaction. The enthalpy of formation of a substance in its standard state, i.e. its stable phase at the standard pressure and temperature, corresponds to its enthalpy compared to those of its constituent elements in their reference states. By convention, the standard reference enthalpy for the elements ($C_{(graphite)}$, O_2, N_2, ...) is zero. The enthalpies of formation $\Delta_f H^0$ that are provided in reference tables are usually given under standard conditions: $p_0 = 101,325$ Pa (1 atm) and $T_0 = 298.15$ K (25°C).

The standard enthalpy $\Delta_r H^0$ of reaction in a reference state can be determined from the standard enthalpies of formation for the reactants r_i and the products p_j, whose stoichiometric coefficients are, respectively, ν_i and μ_j (Figure 4.12).

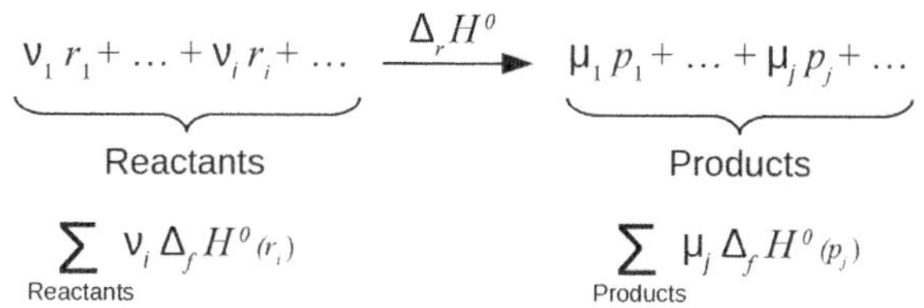

$$\underbrace{\nu_1 r_1 + \ldots + \nu_i r_i + \ldots}_{\text{Reactants}} \xrightarrow{\Delta_r H^0} \underbrace{\mu_1 p_1 + \ldots + \mu_j p_j + \ldots}_{\text{Products}}$$

$$\underbrace{\sum \nu_i \Delta_f H^0 (r_i)}_{\text{Reactants}} \qquad \underbrace{\sum \mu_j \Delta_f H^0 (p_j)}_{\text{Products}}$$

Figure 4.12. *Standard enthalpy of reaction*

The standard enthalpy of reaction is given by:

$$\Delta_r H^0 = \sum_{products} \mu_j \Delta_f H^0(p_j) - \sum_{ractants} \nu_i \Delta_f H^0(r_i) \qquad [4.17]$$

Assuming that the reactant r_1 corresponds to the fuel, the heating value at constant pressure under standard conditions is:

$$HV_p = \frac{|\Delta_r H^0|}{\nu_1 M(r_1)} \qquad [4.18]$$

where $M(r_1)$ is the molar mass of the fuel.

Table 4.3 shows some values of standard molar enthalpies of formation. The indices $(s), (l), (g)$ indicate whether the considered species is in the solid, liquid or gas state, respectively.

Standard enthalpy of formation $\Delta_f H^0$ [kJ/mol] Standard state: $T_0 = 298.15$ K (25°C), $p_0 = 101,325$ Pa (1 atm)						
$C_{(s)}$	$CO_{(g)}$	$CO_{2(g)}$	$H_2O_{(l)}$	$H_2O_{(g)}$	$NO_{(g)}$	$NO_{2(g)}$
0	-110.5	-393.5	-285.8	-241.8	+90.4	+33.8
CH_4 (g)	$n\,C_2H_6$ (g)	$n\,C_4H_{10}$ (g)	C_6H_6 (l)	$n\,C_8H_{18}$ (l)	$n\,C_8H_{18}$ (g)	$n\,C_{12}H_{26}$ (l)
-74.9	-84.7	-393.5	+49.1	-249.7	-208.7	-352.1

Table 4.3. *Standard molar enthalpies of formation*

For example, the combustion of methane is:

$$CH_{4(g)} + 2\,O_{2(g)} \longrightarrow CO_{2(g)} + 2\,H_2O_{(g)} \qquad [4.19]$$

and:

$$\Delta_r H^0 = -393.5 + 2\,(-241.8) - (-74.9) = -802.2 \text{ kJ/mol} \qquad [4.20]$$

for which the standard enthalpy of reaction is *negative*, corresponding to reaction for which energy is transferred from the combustible mixture to the exterior (*exothermic* reaction).

Assuming that the water formed from this reaction is in its vapor state, the lower calorific value at constant pressure is:

$$LHV_p = \frac{802.2}{(12.0107 + 2 \times 2.016)\,10^{-3}} = 50,004 \text{ kJ/kg} \qquad [4.21]$$

4.3.3.2. Lower and upper heating values

As already indicated, the difference between the upper and lower heating values corresponds to the latent heat of vaporization for the water formed by the combustion at the temperature T_0 ($l_v = 2,441.7$ kJ/kg at the temperature $T_0 = 25°C$). The difference between the standard enthalpies of formation of water as a liquid and as a vapor is expressed as:

$$\Delta_f H^0_{(H_2O_{(l)})} - \Delta_f H^0_{(H_2O_{(g)})} = M_{(H_2O)} \, l_{v(T=T_0)}$$
$$= 18.016 \times 10^{-3} \times 2\,441.7 = 44.0 \text{ kJ/mol}$$
[4.22]

with this difference corresponding to the values between the standard enthalpies of formation of water vapor and liquid water, as given in Table 4.3.

Returning now to the example with methane, the higher heating value is obtained by replacing $\Delta_f H^0_{(H_2O_{(g)})}$ by $\Delta_f H^0_{(H_2O_{(l)})}$ in equations [4.20] and [4.21]: $HHV_{p(CH_4)} = 55,489$ kJ/kg.

4.3.3.3. Heating values at a constant pressure and volume

The heating values at a constant pressure and volume were defined by considering the reactions of the combustible mixture initially with the reference pressure p_0 and temperature T_0 conditions, while, respectively, keeping the pressure or volume at a constant value, before bringing the combustion products to their final temperature T_0 (Figure 4.13). During these reactions, the variations in the enthalpy and internal energy are $\Delta_r H^0$ and $\Delta_r U^0$.

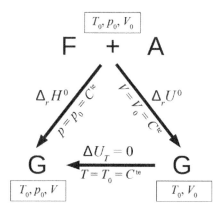

Figure 4.13. *Combustion at constant pressure and volume*

The heating values at a constant pressure HV_p and volume HV_v defined via the reference temperature and reference stoichiometry are given by:

$$HV_p = \frac{|\Delta_r H^0|}{m_F} \qquad HV_v = \frac{|\Delta_r U^0|}{m_F} \qquad [4.23]$$

where m_F is the stoichiometric mass of the fuel used in the combustion.

Since the temperature of the final state of combustion at a constant volume is equal to that of the final state of the reaction at a constant pressure, it is possible to pass from one final state to another by an isothermal transformation at the temperature T_0. If we assume that the combustion gases are either ideal or semi-ideal gases, then their internal energy depends only on the temperature.

During the transformation which brings the combustion gases from the final state with a constant volume to that with a constant pressure, the variation in the internal energy ΔU_T is zero, and:

$$\Delta_r H^0 = \Delta_r U^0 + p_0 \left(V_{(G)} - V_0\right)_{(T=T_0)} = \Delta_r U^0 + \Delta n \, R \, T_0 \qquad [4.24]$$

where Δn is the variation of the number of moles of gas between the initial and final states.

Considering again the case of methane combusting (equation [4.19]), the variation Δn in the number of moles is 1/2 and:

$$\Delta_r U^0 = -802.2 - \frac{1}{2} \times 8.314 \cdot 10^{-3} \times 288 = 803.4 \text{ kJ/mol} \qquad [4.25]$$

which leads to the lower heating value at a constant volume of $PCI_v = 50,079$ kJ/mol. The relative difference between the heating values of methane at a constant pressure and volume is very small, which is generally the case (Sawerysyn 1993).

Heating values can also be expressed in kJ/mol, per unit mass of fuel (kJ/kg or MJ/kg) or per unit volume of fuel (kJ/m^3 or kJ/l). Table 4.4 shows the lower heating values of some fuels.

4.3.4. *Adiabatic flame and explosion temperatures*

To simplify the presentation, we will determine the adiabatic flame and explosion temperatures in the case where methane undergoes combustion in the presence of air. In this case, equations [4.8] (lean mixture – $\lambda > 1$) and [4.11] (rich mixture – $\lambda < 1$) are written respectively as:

$$\lambda \geq 1: \quad CH_4 + 2\lambda(O_2 + 3.76\, N_2) \longrightarrow$$
$$CO_2 + 2\, H_2O + 2\,(\lambda-1)\, O_2 + 2 \times 3.76\, \lambda\, N_2 \qquad [4.26]$$

$$\lambda \leq 1: \quad CH_4 + 2\lambda(O_2 + 3.76\, N_2) \longrightarrow$$
$$(1 - 4(1-\lambda))\, CO_2 + 4(1-\lambda)\, CO + 2\, H_2O + 2 \times 3.76\, \lambda\, N_2$$
[4.27]

where the reactants and combustion gases are assumed to be gaseous.

Heating values at less than 25°C: LHV [MJ/kg]			
Hydrogen	12.10	n-Hexane	44.75
Methane	50.03	n-Heptane	44.57
n-Ethane	47.79	n-Octane	44.43
n-Propane	46.36	n-Nonane	44.31
n-Butane	45.75	n-Undecane	44.19
n-Pentane	45.36	n-Dodecane	44.15
Isobutane	45.61	Isopentane	45.24
Cyclopentane	44.64	Cyclohexane	43,45
Ethylene	47.20	p-Xylene	40.80
Isobutene	45.06	Acetylene	48.24
Benzene	40.17	Toluene	40.59
Methanol	19.94	Ehanol	28.87
Fuel (SP95)	42.9	Natural gas for vehicles (NGV)	48.5
Diesel	42.6		
Groningen Natural gas	40.3	Liquefied Petrol gas (LPG)	46.0
Distributed fuels are mixtures of many chemical species, and their heating values depend on their compositions.			

Table 4.4. *Lower heating values*

Under standard conditions, the enthalpies $\Delta_r H^0_{\to CO_2}$ and $\Delta_r H^0_{\to CO}$ of the carbon dioxide and carbon monoxide produced from the combustion of methane can be obtained from the standard enthalpies of formation (Table 4.3):

$$CH_{4(g)} + 2\, O_{2(g)} \longrightarrow CO_{2(g)} + 2\, H_2O_{(g)}$$
$$\Delta_r H^0_{\to CO_2} = -802.2 \text{ kJ/mol}$$
[4.28]

$$CH_{4(g)} + \frac{3}{2}\, O_{2(g)} \longrightarrow CO_{(g)} + 2\, H_2O_{(g)}$$
$$\Delta_r H^0_{\to CO} = -519.2 \text{ kJ/mol}$$
[4.29]

152 Thermodynamics of Heat Engines

Assuming that the combustible mixture is at the standard conditions initially, the *adiabatic flame temperature* T_f is the temperature of the gases produced by the adiabatic combustion of the mixture at a constant pressure. According to the first law of thermodynamics, the enthalpy variation during this evolution is zero ($\Delta H = 0$).

In the case of a *lean mixture*, the combustion reaction can be broken down into two parts (Figure 4.14):

– the species that are involved in chemical transformations: $CH_4 + 2\,O_2$;

– the species that are present but do not undergo any chemical reactions: the excess oxygen $2\,(\lambda - 1)\,O_2$ and nitrogen from the air $2 \times 3.76\,\lambda\,N_2$.

Since enthalpy being a state quantity, its variation between the initial and the final states ΔH_R of the species participating in the chemical reaction can be expressed in the form:

$$\Delta H_R = \Delta_r H^0_{\to CO_2} + \Delta H_P \qquad [4.30]$$

and, as the enthalpy is an extensive quantity:

$$\Delta H_P = \Delta H_{CO_2(T_0 \to T_f)} + 2\,\Delta H_{H_2O(T_0 \to T_f)} \qquad [4.31]$$

Similarly, the enthalpy variation ΔH_N of a species that does not undergo any chemical reaction is expressed in the form:

$$\Delta H_N = 2\,(\lambda - 1)\,\Delta H_{O_2(T_0 \to T_f)} + 2 \times 3.76\,\lambda\,\Delta H_{N_2(T_0 \to T_f)} \qquad [4.32]$$

$\Delta H_{CO_2(T_0 \to T_f)}$, $\Delta H_{H_2O(T_0 \to T_f)}$, $\Delta H_{O_2(T_0 \to T_f)}$, $\Delta H_{N_2(T_0 \to T_f)}$ are the molar enthalpy variations of carbon dioxide CO_2, water vapor H_2O, oxygen O_2, and nitrogen N_2, respectively, when heated at a constant pressure p_0 from the initial temperature T_0 to the final temperature T_f.

According to the first law of thermodynamics:

$$\Delta H = \Delta H_R + \Delta H_N = \Delta_r H^0_{\to CO_2} + \Delta H_P + \Delta H_N = 0 \qquad [4.33]$$

and the equation, from which we can calculate the flame temperature, can be deduced from it:

$$\Delta H_{CO_2(T_0 \to T_f)} + 2\,\Delta H_{H_2O(T_0 \to T_f)} + 2\,(\lambda - 1)\,\Delta H_{O_2(T_0 \to T_f)}$$
$$+\,2 \times 3.76\,\lambda\,\Delta H_{N_2(T_0 \to T_f)} = -\Delta_r H^0_{\to CO_2} \qquad [4.34]$$

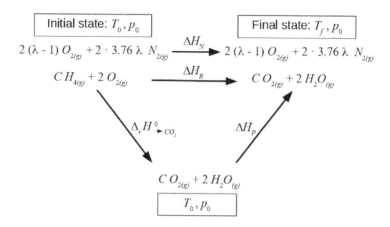

Figure 4.14. *Flame temperature for a lean methane–air mixture*

If the species A_i that constitute the air and combustion gases are considered to be semi-ideal gases, the molar enthalpy variations $\Delta H_{A_i(T_0 \rightarrow T_f)}$ which correspond to the temperature variation from T_0 to T_f at a constant pressure can be expressed by:

$$\Delta H_{A_i(T_0 \rightarrow T_f)} = \int_{T_0}^{T_f} C_{p(A_i)}(T)\, dT \qquad [4.35]$$

where $C_{p(A_i)}$ is the molar heat capacity at a constant pressure of the species A_i, which is only a function of the temperature T.

In the case of a *rich* methane–air mixture, Figure 4.15 illustrates the decomposition of equation [4.26] into its reactive and neutral components.

The equation that defines the flame temperature of the rich mixture is:

$$(1 - 4(1-\lambda))\Delta H_{CO_2(T_0 \rightarrow T_f)} + 4(1-\lambda)\Delta H_{CO(T_0 \rightarrow T_f)}$$
$$+ 2\Delta H_{H_2O(T_0 \rightarrow T_f)} + 2 \times 3.76\, \lambda\, \Delta H_{N_2(T_0 \rightarrow T_f)}$$
$$= -(1 - 4(1-\lambda))\Delta_r H^0_{\rightarrow CO_2} - 4(1-\lambda)\Delta_r H^0_{\rightarrow CO} \qquad [4.36]$$

Figure 4.16 represents the behavior of the flame temperature for the combustion of methane with air as the oxidant, as a function of the air–fuel equivalence ratio calculated using equations [4.34] and [4.36].

The molar thermal capacities at a constant pressure for the different species are estimated using polynomial representations (equation [1.60]), by using the coefficients in Table 1.2. The flame temperature is the greatest in the case of the stoichiometric mixture ($\lambda = 1$). The decay of the flame temperature for a rich mixture ($\lambda < 1$) when the air–fuel equivalence ratio decreases comes from the incomplete combustion of methane, which releases less energy than in the case of complete oxidation ($\Delta_r H^0_{\to CO} < \Delta_r H^0_{\to CO_2}$). In the case of a poor mixture ($\lambda > 1$), the energy released from the combustion remains constant. The energy needed to heat up excess gases explains the decrease in the flame temperature when the air–fuel equivalence ratio increases.

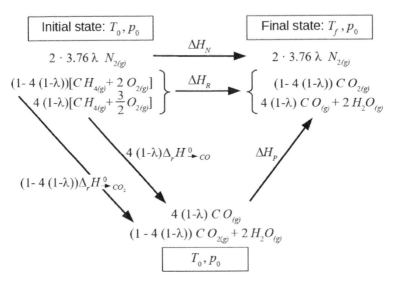

Figure 4.15. *Flame temperature of a rich methane–air mixture*

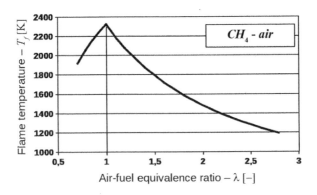

Figure 4.16. *Flame temperature of the methane–air mixture as a function of the air factor*

Assuming that the combustible mixture is initially at the standard conditions, the *explosion adiabatic temperature* T_E is the temperature of the gases produced by the constant volume adiabatic combustion of the mixture. The temperature calculation T_E can be derived analogously to that of the flame temperature, by replacing enthalpies by internal energies.

4.4. Chemical kinetics of combustion

4.4.1. *Chain reactions*

The reaction for a theoretical combustion with an excess of air [4.8] presumes that CO_2 and H_2O, in addition to excess oxygen and air nitrogen, are products of the reaction. In the real case of hydrocarbon undergoing combustion, *dissociations*, which are especially important at temperatures above 1,200 °C, arise when in the presence of appreciable quantities of CO, H_2, O_2 and come from the equilibrium reactions:

$$2\,CO_2 \rightleftharpoons 2\,CO + O_2$$
$$2\,H_2O \rightleftharpoons 2\,H_2 + O_2$$
[4.37]

In combustion reactions, many *radicals* (chemical species with one or more non-paired electrons in its outer shell) are generated: $OH^\bullet$, $^\bullet O$, $H^\bullet$, $NO^\bullet$, ..., which generally are very unstable and are likely to react with many compounds.

The formation of the combustion products comes from a complex mechanism of chain reactions, that can be decomposed generally into three steps (Borghi and Destriau 1995; Griffiths and Barnard 1995):

– *The initiation step*: the first reaction of the chain, for example:

$$H_2 + O_2 \rightarrow 2\,OH^\bullet \qquad [4.38]$$

– *The propagation step*:

- Chain propagation: a radical on the left produces one on the right. For example:

$$OH^\bullet + H_2 \rightarrow H_2O + H^\bullet \qquad [4.39]$$

- Chain branching: a radical on the left produces two on the right. For example:

$$H^\bullet + O_2 \rightarrow OH^\bullet + {}^\bullet O$$
$$O^\bullet + H_2 \rightarrow OH^\bullet + H^\bullet$$
[4.40]

– *The termination step*:

$$H^\bullet + H^\bullet + M \rightarrow H_2 + M$$
$$H^\bullet + OH^\bullet + M \rightarrow H_2O + M$$
[4.41]

The third species is denoted M, which can be the wall for example, which stabilizes the reaction by decreasing the kinetic energy by collisions.

The mechanisms of a chain reaction play a vital role with regard to the formation of pollutants too.

Because of the thermal exchanges and the endothermic dissociation reactions, the temperature of real combustion is significantly lower than that determined from the theoretical combustion equation.

The purpose of chemical kinetics is *to describe the composition of a system as a function of time*, whose constituents are reacting between themselves.

4.4.2. Composition of a reactive mixture

The composition of a reactive mixture constituting n moles $A_i (i = 1, ..., n)$ can be characterized by:

– the *molar concentrations* of the different constituents:

$$C_{A_i} = [A_i] = \frac{n_i}{V}$$
[4.42]

where n_i is the number of constituent moles A_i, and V is the volume occupied by the mixture. Molar concentrations can be expressed in mol/m³, mol/L or mol/cm³;

– the *molar fractions (or titers)*:

$$y_i = \frac{n_i}{n} \quad \text{with:} \quad \sum_{i=1}^{n} n_i = n \implies \sum_{i=1}^{n} y_i = 1$$
[4.43]

– the *partial pressures*, in the case of gases: The partial pressures p_i of the constituent gases A_i is the pressure that the constituent gas would have if it solely occupied the volume V within the mixture at a temperature equilibrium. For ideal or semi-ideal gases:

$$p_i V = n_i RT \quad \text{with:} \quad \sum_{i=1}^{n} p_i = p$$
[4.44]

where p is the pressure of the mixture at temperature T.

According to the state equation, we obtain:

$$C_{A_i} = [A_i] = \frac{p_i}{RT} \quad \text{and:} \quad y_i = \frac{p_i}{p} \qquad [4.45]$$

4.4.3. Reaction rates

The reaction which we are considering, whose reactants A_i generate the products A'_i, can be written in the form:

$$\nu_1 A_1 + \nu_2 A_2 + \ldots + \nu_i A_i + \ldots \rightarrow \nu'_1 A'_1 + \ldots + \nu'_j A'_j + \ldots \qquad [4.46]$$

where the ν_i and ν'_j are the stoichiometric coefficients of the reaction.

The *extent of reaction* at the time t is defined by the relation:

$$\xi(t) = \frac{n_i(t) - n_{i0}}{(-\nu_i)} = \frac{n'_j(t) - n'_{j0}}{\nu'_j} \qquad [4.47]$$

The n_{i0} and n'_{j0} are the initial values of the numbers of moles n_i and n'_j, respectively. The definition of $\xi(t)$ is the same as expressing the number of moles of the different reactants as being proportional to their stoichiometric coefficients (Figure 4.17).

$\xi(t)$ is an extensive physical quantity.

	$\nu_1 A_1 + \ldots + \nu_i A_i + \ldots$		$\rightarrow \nu'_1 A'_1 + \ldots + \nu'_j A'_j + \ldots$	
$t = 0:$	n_{10}	n_{i0}	n'_{10}	n'_{j0}
$t:$	$n_{10} - \nu_1 \xi(t)$	$n_{i0} - \nu_i \xi(t)$	$n'_{10} + \nu'_1 \xi(t)$	$n'_{j0} + \nu'_j \xi(t)$

Figure 4.17. *Extent of a chemical reaction*

The *molar rate of reaction* or *rate of conversion*, which is an extensive quantity, is defined by the relation:

$$\frac{d\xi}{dt} = -\frac{1}{\nu_i}\frac{dn_i}{dt} = \frac{1}{\nu'_j}\frac{dn'_j}{dt} \qquad [4.48]$$

Dividing the previous expression by the unit volume, we obtain the *specific rate* of the chemical reaction w, which is an intensive quantity:

$$w = \frac{1}{V}\frac{d\xi}{dt} = -\frac{1}{\nu_i V}\frac{dn_i}{dt} = \frac{1}{\nu'_j V}\frac{dn'_j}{dt} \qquad [4.49]$$

From the relation given by definition [4.42]:

$$\frac{d[A_i]}{dt} = \frac{d}{dt}\left(\frac{n_i}{V}\right) = \frac{1}{V}\frac{dn_i}{dt} - \frac{[A_i]}{V}\frac{dV}{dt} \quad [4.50]$$

If the reaction is isochoric ($V = C^{te}$), or if the volume changes very slowly with time ($[A_i]\, dV/dt \ll dn_i/dt$), the specific rate w can be expressed as follows:

$$w = -\frac{1}{\nu_i}\frac{d[A_i]}{dt} = \frac{1}{\nu'_j}\frac{d[A'_j]}{dt} \quad [4.51]$$

Many reactions have a kinetic equation of the form:

$$w(t) = k_c\, [A_1]_{(t)}^{q_1}\, [A_2]_{(t)}^{q_2}\, \ldots = k_c \prod_i [A_i]_{(t)}^{q_i} \quad [4.52]$$

where the constants q_i are the *partial reaction orders* of the reaction with respect to the constituents A_i, and $q = \sum_i q_i$ is the *overall reaction order* of the reaction. k_c is the rate "constant", whose dimension, which depends on the overall reaction order, is generally expressed in $\text{mol}^{1-q} \cdot \text{cm}^{3\,q-3} \cdot \text{s}^{-1}$.

In the particular case when the partial order is equal to the stoichiometric coefficient for any reactant (Van't Hoff's law), the reaction is said to be of simple order. Then:

$$\forall i\ \ q_i = \nu_i \implies w(t) = k_c \prod_i [A_i]_{(t)}^{\nu_i} \quad [4.53]$$

Experience shows that the reaction rate "constant" k_c depends on the temperature. It is often represented as:

$$k_c = A\, T^n\, e^{-\frac{B}{T}} \quad [4.54]$$

Here, A is the *pre-exponential factor*, or the *frequency factor* and, in setting $B = \frac{E_a}{R}$:

$$k_c = A\, T^n\, e^{-\frac{E_a}{RT}} \quad [4.55]$$

$B = E_a/R = T_a$ has the dimension of temperature (where T_a is the activation temperature) and E_a is defined to be the "*activation energy*".

The activation energy E_a can be interpreted as the required energy to be supplied to the chemical system for the reaction to occur. The pre-exponential factor A is proportional to the collision frequency between molecules, taking into account the positional effects of the atoms making up the molecules.

In the particular case of $n = 0$, $k_c(T)$ obeys the Arrhenius equation:

$$k_c = A\, e^{-\frac{E_a}{RT}} \quad [4.56]$$

Using relations [4.43] and [4.45], equation [4.53] can be expressed in terms of the molar fractions y_i:

$$w(t) = k_y \prod_i y_i^{q_i} \quad \text{with:} \quad k_y = k_c \left(\frac{n}{V}\right)^q \qquad [4.57]$$

and, in the case of a reactive mixture of ideal or semi-ideal gases, then from relation [4.45]:

$$w(t) = k_p \prod_i p_i^{q_i} \quad \text{with:} \quad k_p = \frac{k_c}{(RT)^q} \qquad [4.58]$$

4.4.4. Establishing a chemical equilibrium

Many chemical reactions are *reversible* (Figure 4.18). For the direction of the reaction defined as being the forward one (from left to right), q_i is the order of the reaction for the component A_i, and k_c^+ is the rate constant. For the reverse reaction, q'_j is the order of the reaction with respect to the component A'_j, and k_c^- is the rate constant.

$$\nu_1 A_1 + \ldots + \nu_i A_i + \ldots \underset{k_c^-}{\overset{k_c^+}{\rightleftharpoons}} \nu'_1 A'_1 + \ldots + \nu'_j A'_j + \ldots$$

Figure 4.18. *A reversible chemical reaction*

Relation [4.59] shows the contribution of the direct reaction to the destruction of $[A_i]$ species and the production of $[A'_j]$ species:

$$w^+(t) = -\frac{1}{\nu_i}\left(\frac{d[A_i]}{dt}\right)^+ = \frac{1}{\nu'_j}\left(\frac{d[A'_j]}{dt}\right)^+ = k_c^+ \prod_i [A_i]^{q_i} \qquad [4.59]$$

Similarly, relation [4.60] shows the contribution of the reverse reaction to the destruction of $[A'_j]$ species and the production of $[A_i]$ species:

$$w^-(t) = \frac{1}{\nu_i}\left(\frac{d[A_i]}{dt}\right)^- = -\frac{1}{\nu'_j}\left(\frac{d[A'_j]}{dt}\right)^- = k_c^- \prod_j [A'_j]^{q'_j} \qquad [4.60]$$

The net productions or destructions of the different constituents result from the opposite effects of direct and reverse reactions:

$$\frac{d[A_i]}{dt} = \left(\frac{d[A_i]}{dt}\right)^+ + \left(\frac{d[A_i]}{dt}\right)^- \qquad [4.61]$$

By transferring relations [4.59] and [4.60] into relation [4.61]:

$$-\frac{1}{\nu_i}\frac{d[A_i]}{dt} = k_c^+ \prod_i [A_i]^{q_i} - k_c^- \prod_j [A'_j]^{q'_j} \qquad [4.62]$$

Under the conditions of *chemical equilibrium*, the concentrations of the different constituents remain constant, so:

$$\frac{d[A_i]}{dt} = 0 \implies \frac{\prod_j [A'_j]^{q'_j}}{\prod_i [A_i]^{q_i}} = \frac{k_c^+(T)}{k_c^-(T)} = K_c(T) \qquad [4.63]$$

where $K_c(T)$ is the equilibrium constant.

Van't Hoff's relation allows us to determine the equilibrium constant at different temperatures T, provided we know the standard enthalpy of reaction $\Delta_r H^0$ for the reaction:

$$\frac{d\ln K_c(T)}{dT} = \frac{\Delta_r H^0}{R\,T^2} \qquad [4.64]$$

where R is the molar ideal gas constant.

In the case of ideal or semi-ideal gases, and similar to what was done in the previous chapter, equation [4.63] can be expressed in terms of partial pressures or molar fractions:

$$\frac{\prod_j [p'_j]^{q'_j}}{\prod_i [p_i]^{q_i}} = (R\,T)^{(q-q')}\,K_c(T) = K_p(T) \qquad [4.65]$$

$$\frac{\prod_j [y'_j]^{q'_j}}{\prod_i [y_i]^{q_i}} = p^{(q'-q)}\,K_p(T) \qquad [4.66]$$

4.4.5. Equilibrium composition of the combustion products

At high temperatures, the combustion product compositions differ from those which are predicted from the theoretical chemical reactions. Under equilibrium conditions, these differences are explained by the dissociations which occur at high temperatures. At high temperatures, the equilibrium conditions generally provide a good estimate of what the composition of the products will be.

In the case of a *chemical equilibrium that respects Van't Hoff's law*, and whose reactants and products are either ideal or semi-ideal gases:

$$\ldots + \nu_i\,A_i + \ldots \rightleftharpoons \ldots + \nu'_j\,A'_j + \ldots \qquad [4.67]$$

and, by expressing the concentrations $[A_i]$ in terms of their molar fractions y_i:

$$[A_i] = y_i \left(\frac{p_0}{RT}\right)\left(\frac{p}{p_0}\right) \qquad [4.68]$$

where p is the pressure of the reactant mixture, and p_0 is a pressure chosen to be the reference which is most often the usual atmospheric pressure ($p_0 = 101,325$ Pa), equation [4.63] can be written as:

$$\frac{\prod_j [y'_j]^{\nu'_j}}{\prod_i [y_i]^{\nu_i}} = \left(\frac{RT}{p_0}\right)^{\sum_j \nu'_j - \sum_i \nu_i} K_c(T) \left(\frac{p}{p_0}\right)^{\sum_i \nu_i - \sum_j \nu'_j}$$

$$= K(T) \left(\frac{p}{p_0}\right)^{\sum_i \nu_i - \sum_j \nu'_j} \qquad [4.69]$$

The hydrogen–oxygen combustion reaction with stoichiometric proportions is used as an example, to illustrate the calculation of the equilibrium composition of reactant mixtures concretely. The *theoretical chemical reaction* has only water as the combustion product:

$$2\,H_2 + O_2 \rightarrow 2\,H_2O \qquad [4.70]$$

The phenomenon of *high temperature dissociations* explains the presence of H_2 and O_2 appearing as combustion products, as well as radicals. Only H, O and OH are considered here. The composition of the combustion products can then be written as follows:

$$n_{H_2O}\,H_2O + n_{H_2}\,H_2 + n_{O_2}\,O_2 + n_H\,H + n_O\,O + n_{OH}\,OH \qquad [4.71]$$

The conservation of hydrogen and oxygen is expressed as:

$$n_{H_2O} + n_{H_2} + \frac{n_H}{2} + \frac{n_{OH}}{2} = 2$$

$$\frac{n_{H_2O}}{2} + n_{O_2} + \frac{n_O}{2} + \frac{n_{OH}}{2} = 1 \qquad [4.72]$$

Dividing the two equations [4.72] by the total number of moles n of the products, and relating the resulting expressions:

$$y_{H_2} + \frac{y_H}{2} = 2\,y_{O_2} + y_O + \frac{y_{OH}}{2} \qquad [4.73]$$

where the y_i are the molar fractions of the combustion products, and:

$$y_{H_2O} + y_{H_2} + y_{O_2} + y_H + y_O + y_{OH} = 1 \qquad [4.74]$$

To calculate the six molar fractions, the calculation of equations [4.73] and [4.74] with four independent equations is required. These are obtained from four equilibrium relations, whose constants $K(T)$ (relation [4.69]) are known. Using the equilibrium whose table of values for the constants K are available in reference (Borgnakke and Sonntag 2012), we find:

$$(1): H_2 \rightleftharpoons 2H \qquad \frac{y_H^2}{y_{H_2}} = K_1(T)\left(\frac{p}{p_0}\right)^{-1} \qquad [4.75]$$

$$(2): O_2 \rightleftharpoons 2O \qquad \frac{y_O^2}{y_{O_2}} = K_2(T)\left(\frac{p}{p_0}\right)^{-1} \qquad [4.76]$$

$$(3): 2H_2O \rightleftharpoons 2H_2 + O_2 \qquad \frac{y_{H_2}^2 \, y_{O_2}}{y_{H_2O}^2} = K_3(T)\left(\frac{p}{p_0}\right)^{-1} \qquad [4.77]$$

$$(4): 2H_2O \rightleftharpoons H_2 + 2OH \qquad \frac{y_{H_2}\, y_{OH}^2}{y_{H_2O}^2} = K_4(T)\left(\frac{p}{p_0}\right)^{-1} \qquad [4.78]$$

Figure 4.19 shows that the values from the equilibrium constants tables in the temperature range of 1,500 to 4,000 K are represented with good accuracy by the Arrhenius equation whose coefficients are listed in Table 4.5.

$K(T) = A \exp\left(-\dfrac{B}{T}\right)$ Reference pressure: $p_0 = 101,325$ Pa	A [-]	B [K]
$H_2 \rightleftharpoons 2H$	2.2570×10^6	54 901
$O_2 \rightleftharpoons 2O$	1.0473×10^7	61 550
$N_2 \rightleftharpoons 2N$	1.0988×10^7	115 709
$2H_2O \rightleftharpoons 2H_2 + O_2$	1.2509×10^6	60 659
$2H_2O \rightleftharpoons H_2 + 2OH$	3.1482×10^7	69 336
$2CO_2 \rightleftharpoons 2CO + O_2$	4.2834×10^8	66 332
$N_2 + O_2 \rightleftharpoons 2NO$	2.0653×10^1	21 709
$N_2 + 2O_2 \rightleftharpoons 2NO_2$	2.9531×10^{-7}	8 188.5

Table 4.5. *Equilibrium constants table*

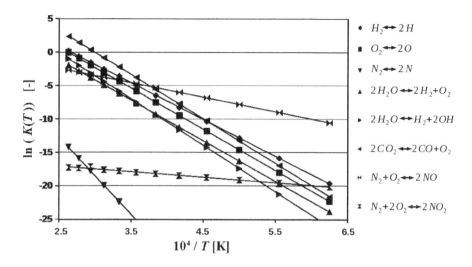

Figure 4.19. *Equilibrium constants as a function of temperature*

Figure 4.20 represents the results obtained via solving by successive approximations for the solutions of equations [4.73]–[4.78]. The curves represent the molar fractions as a function of temperature.

When the pressure is equal to standard atmospheric pressure $p_0 = 101,325$ Pa (Figure 4.20(a)), the molar fraction of water remains very close to unity at temperatures below 2,000 K, indicating that dissociations remain very low. At higher temperatures, as shown in the figure, the molar fraction of water decreases and the molar fractions of the dissociated species become significant, resulting in a reduction in the amount of heat supplied by the combustion, as dissociation is an endothermic phenomenon. In the presence of dissociations, in the case where the aim is to calculate the combustion temperatures, it is necessary to couple the calculation of the reactant mixture composition with the energy balance.

Figure 4.20 (b) compares the molar fractions when the pressures are p_0 and $10 \times p_0$, respectively. It is clear that the dissociations become less pronounced as the pressure increases.

The model with four elementary chemical equations presented in this chapter is a simplified model that does not take into account some species, such as HO_2 and H_2O_2. Boivin et al. (2011) cites the so-called San Diego mechanism, which is known to give good estimates of the product composition from the hydrogen–oxygen combustion, and which involves 21 elementary chemical reactions.

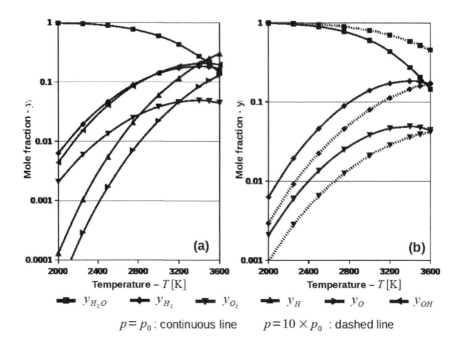

Figure 4.20. *Molar fractions as a function of temperature*

4.4.6. *Detailed chemical kinetics–formation of pollutants*

The objective of detailed chemical kinetics is to describe the transformation of reactants into products at the molecular level. It is based on chemical mechanisms, consisting of a number of elementary reactions, whose rate constants as a function of temperature are known.

The combustion of hydrocarbons using air as an oxidant is a particularly complex process. The fuels used in both land and air transport are mixtures of a large number of hydrocarbons, and their oxidation reactions involve a large number of steps with several intermediate reaction. These partial elementary reactions are at the origin of the formation of *polluting species*, in particular *carbon monoxide* CO and *nitrogen oxides* NO, NO_2 and N_2O, often grouped together under the term NOx.

A detailed chemical mechanism is constructed by considering the intermediate species involved in the combustion phenomenon studied, and by associating all the elementary reactions in which these species are involved.

For example, in the case of methane combustion (CH_4), Miyamoto et al. (1990) considers 15 chemical species ($CH_4, CH_3, CH_2O, CHO, CO, CO_2, H_2O$,

$H_2, O_2, OH, H, O, N_2, NO, N$) and 31 elementary chemical equations. Still considering methane combustion, Bibrzycki and Poinsot (2011) cites the "GRI Mech 3.0" model with 52 species and 323 elementary chemical equations. The oxidation of more complex hydrocarbons requires the consideration of an even greater number of intermediate species and chemical equations: Titova et al. (2011) develops a model that describes the oxidation of n-decane, with 144 species and 1,021 chemical reactions.

The systems of differential equations that are deduced from chemical mechanisms are very "stiff", and the classical methods used to solve ordinary differential equations explicitly (Runge–Kutta, etc.) are in general not very successful; it is therefore necessary to use specific implicit methods.

It is often useful to use combustion models that require less computing power than detailed models. Reduced models can be defined using the sensitivity analysis of the different elementary reactions. It is recognized that a good four-step reduced model has been proposed by Jones and Linstedt (1988), for the combustion of alkanes up to butane, using air as the oxidant in premixed and diffusion flames (Figure 4.21).

$$
\begin{align}
(1) \quad & C_n H_{2n+2} + \frac{n}{2} O_2 \longrightarrow n\, CO + (n+1)\, H_2 \\
(2) \quad & C_n H_{2n+2} + n\, H_2O \longrightarrow n\, CO + (2n+1)\, H_2 \\
(3) \quad & H_2 + \frac{1}{2} O_2 \rightleftharpoons H_2O \\
(4) \quad & CO + H_2O \rightleftharpoons CO_2 + H_2
\end{align}
$$

Figure 4.21. *Reduced chemical mechanism of Jones and Linstedt (1988)*

It should be noted that carbon monoxide (CO), which is one of the main polluting combustion products, appears in the Jones and Linstedt mechanism. This mechanism is used to estimate the concentration of carbon monoxide in the combustion products.

At *high temperatures*, of about 1,500°C, the thermal energy is sufficient to cause oxygen and nitrogen to dissociate, which then recombine to form nitrogen oxides. The formation of nitrogen oxides at high temperatures ("thermal NO") is generally described by the Zeldovitch et al. (1947), shown in Figure 4.22, which allows for the concentration of nitrogen oxide to be expressed as:

$$\frac{d[NO]}{dt} = k_{c1}^+ [O][N_2] + k_{c2}^+ [N][O_2] + k_{c3}^+ [N][OH]$$
$$- k_{c1}^- [NO][N] - k_{c2}^- [NO][O] - k_{c3}^- [NO][H]$$

[4.79]

$$
\begin{aligned}
&(1) \quad O + N_2 \underset{k_{c1}^-}{\overset{k_{c1}^+}{\rightleftharpoons}} NO + N \\
&(2) \quad N + O_2 \underset{k_{c2}^-}{\overset{k_{c2}^+}{\rightleftharpoons}} NO + O \\
&(3) \quad N + OH \underset{k_{c3}^-}{\overset{k_{c3}^+}{\rightleftharpoons}} NO + H
\end{aligned}
$$

Figure 4.22. *Zeldovich's mechanism (Zeldovitch et al. 1947)*

A complementary pathway for the formation of nitrogen oxides ("*NO prompt*") was introduced by Fenimore (1971). Fenimore proposes (Figure 4.23) that CH radicals react with N_2 nitrogen to form the reaction intermediates HCN and N (reaction (1)), which are then oxidised to form the nitrogen oxides.

$$(1) \quad CH + N_2 \rightleftharpoons HCN + N$$

Reactions with N:

(2) $N + O_2 \longrightarrow NO + O$

(3) $N + NO \longrightarrow N_2 + O$

Reactions with HCN:

(4) $HCN + O \rightleftharpoons NCO + H$

(5) $NCO + H \rightleftharpoons NH + CO$

(6) $NH + H \rightleftharpoons N + H_2$

(7) $N + OH \rightleftharpoons NO + H$

Figure 4.23. *Formation of the NO prompt according to Fenimore (1971)*

In the case where the fuel contains nitrogen, a third pathway for the formation of nitrogen oxides ("*fuel NOx*") is identified. Most of the nitrogen combined with the fuel is found as NOx in the combustion products.

4.5. Exergy analysis of combustion

4.5.1. *Exergy of a gas mixture*

Consider a gas mixture at a pressure p and thermodynamic temperature T, whose N constituents A_i obey the ideal gas equation of state:

$$\text{mixture}: \quad \sum_{i=1}^{N} n_i\, A_i \qquad [4.80]$$

where n_i represents the number of moles of the species A_i.

The exergy of the mixture is defined as:

$$Ex = H - H^0 - T_0 (S - S^0) \qquad [4.81]$$

$H - H^0$ and $S - S^0$ are the enthalpy and entropy variations of the system, between the considered state and the reference state chosen as reference (pressure $p_0 = 0.1$ MPa (1 bar); temperature $T_0 = 298$ K (25°C)).

Since the mixture is made up of gases that obey with the ideal gas equation of state, the enthalpy change $H - H^0$ of the mixture is given by:

$$H - H^0 = \sum_{i=1}^{N} n_i \int_{T_0}^{T} C_{pi}(T)\, dT \qquad [4.82]$$

where $C_{pi}(T)$ is the molar heat capacity at a constant pressure for the species A_i and, according to relation [1.53]:

$$dS_i = C_{pi}(T) \frac{dT}{T} - R \frac{dp_i}{p_i} \qquad [4.83]$$

where S_i and p_i are the molar entropy and the *partial pressure* in the mixture of the species A_i, respectively, and R is the molar ideal gas constant. By integrating equation [4.83] between the reference conditions and the state being considered:

$$S_i - S_i^0 = \int_{T_0}^{T} C_{pi}(T) \frac{dT}{T} - R \ln\left(\frac{p_i}{p_0}\right) \qquad [4.84]$$

with:

$$y_i = \frac{n_i}{\sum_{i=1}^{N} n_i} \qquad [4.85]$$

Given $p_i = y_i\, p$, the entropy difference $S - S^0$ of the gas mixture between the reference and considered states is:

$$S - S^0 = \sum_{i=1}^{N} n_i \left(\int_{T_0}^{T} C_{pi}(T) \frac{dT}{T} - R \ln\left(y_i \frac{p}{p_0}\right) \right) \qquad [4.86]$$

By substituting expressions [4.82] and [4.86] into [4.81]:

$$Ex = \sum_{i=1}^{N} n_i \left[\int_{T_0}^{T} Cpi(T)\, dT - T_0 \left(\int_{T_0}^{T} C_{pi}(T) \frac{dT}{T} - R \ln\left(y_i \frac{p}{p_0}\right) \right) \right]$$

$$\qquad [4.87]$$

REMARK.–

Consider the mixture of two gases A_1 (n_1 moles) and A_2 (n_2 moles), initially under the reference conditions p_0, T_0 (Figure 4.24).

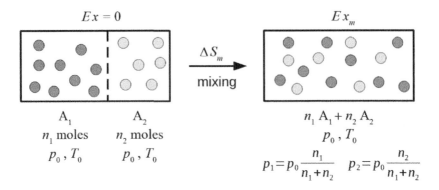

Figure 4.24. *Mixing two gases*

It is assumed that the mixture is formed *at constant pressure and temperature*. In the mixture, the two species are in thermal equilibrium with the reference temperature T_0, and the partial pressures p_1 and p_2 of the species A_1 and A_2 are, respectively:

$$p_1 = p_0 \frac{n_1}{n_1 + n_2} = y_1\, p_0 \qquad p_2 = p_0 \frac{n_2}{n_1 + n_2} = y_2\, p_0 \qquad [4.88]$$

where y_1 and y_2 are the molar fractions.

Relation [4.86] allows us to determine the entropy variation ΔS_m that results from the mixing:

$$\Delta S_m = -n_1\, R\, \ln(y_1) - n_2\, R\, \ln(y_2) = -R\, \ln\left(y_1^{n_1}\, y_2^{n_2}\right) \qquad [4.89]$$

As the molar fractions are quantities less than one, the value of ΔS_m is positive, which expresses the irreversibility of the mixture of the two species.

The exergy for both of the two species is zero for the initial conditions; hence, the exergy change that comes from the mixing and corresponds to the exergy Ex_m with the final conditions is given by:

$$Ex_m = R\, T_0\, \ln\left(y_1^{n_1}\, y_2^{n_2}\right) \qquad [4.90]$$

which is a negative quantity which shows that it is necessary to provide some energy to separate the components of the mixture, in order to return them to their reference states.

4.5.2. *Exergy production from a combustion reaction*

Consider the reaction of a combustible mixture $F + A : \sum_i n_i A_i$, initially at the temperature T_m and pressure p_m and which produces the combustion gases $G : \sum_j n'_j A'_j$ at the temperature T_m and pressure p_m (Figure 4.25). Here, n_i and n'_j are the number of moles of the species A_i and A'_j, respectively, and constitute the reactant mixture and the combustion gases. We wish to determine the exergy increase $\Delta_r Ex$ of the combustion gases, with respect to the exergy of the fuel mixture.

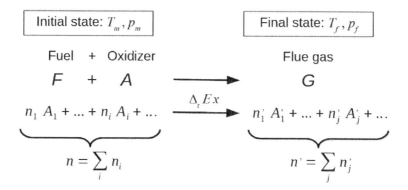

Figure 4.25. *A combustion reaction*

Since the species A_i and A'_j that are considered here to obey the ideal gas equation of state, their enthalpies depend solely on the temperature and the exergies of the combustible mixture Ex_M with the initial conditions and of the combustion gases Ex_G, which can be expressed by:

$$Ex_M = \sum_i n_i \left(H_{A_i}(T_m) - H_{A_i}(T_0) \right)$$

$$- T_0 \left(\sum_i n_i \left(S_{A_i}(T_m, P_m) - S^0_{A_i} \right) \right) \quad [4.91]$$

$$Ex_G = \sum_j n'_j \left(H_{A'_j}(T_f) - H_{A'_j}(T_0) \right)$$

$$- T_0 \left(\sum_j n'_j \left(S_{A'_j}(T_f, P_f) - S^0_{A'_j} \right) \right) \quad [4.92]$$

Given the assumption of the nature of the constituents A_i and A'_j, their molar enthalpies relative to the reference state are:

$$H_{A_i}(T_m) - H_{A_i}(T_0) = \int_{T_0}^{T_m} C_{pA_i}(T)\, dT$$

$$H_{A'_j}(T_f) - H_{A'_j}(T_0) = \int_{T_0}^{T_f} C_{pA'_j}(T)\, dT \qquad [4.93]$$

and entropies relative to the reference state are:

$$S_{A_i}(T_m, p_m) - S^0_{A_i} = \int_{T_0}^{T_m} C_{pA_i}(T) \frac{dT}{T} - R \ln\left(\frac{p_i}{p_0}\right) \qquad [4.94]$$

where R is the molar ideal gas constant and p_i is the partial pressure of the species A_i in the mixture, which can be expressed by introducing the molar fraction y_i of the respective species ($p_i = y_i\, p_m$). The same approach can be applied to the combustion gases, from which we obtain:

$$S_{A_i}(T_m, p_m) - S^0_{A_i} = \int_{T_0}^{T_m} C_{pA_i}(T) \frac{dT}{T} - R \ln\left(y_i \frac{p_m}{p_0}\right)$$

$$S_{A'_j}(T_f, p_f) - S^0_{A'_j} = \int_{T_0}^{T_f} C_{pA'_j}(T) \frac{dT}{T} - R \ln\left(y_j \frac{p_f}{p_0}\right) \qquad [4.95]$$

To demonstrate the use of the above relationships, let us consider the theoretical combustion of hydrogen using air as oxidant (Figure 4.26). The fuel–air mixture is initially under the reference conditions (temperature T_0; pressure p_0). Combustion is carried out without any heat exchange with the outside (adiabatic transformation) and at constant pressure p_0, and T_f is the temperature of the burnt gases after combustion (adiabatic flame temperature). We propose to determine the temperature T_f of the combustion gases and the variation of exergy $\Delta_r Ex$ of the gases, resulting from combustion as a function of the excess air e.

The adiabatic flame temperature f (see section 4.3.4) is obtained by solving the equation:

$$\left(-2\, \Delta_r H^0_{(H_2)}\right) = 2 \int_{T_0}^{T_f} C_{p(H_2O)}(T)\, dT + e \int_{T_0}^{T_f} C_{p(O_2)}(T)\, dT$$

$$+ 3.76\,(1+e) \int_{T_0}^{T_f} C_{p(N_2)}(T)\, dT$$

$$[4.96]$$

where $\Delta_r H^0_{(H_2)} = -241.8$ kJ/mol is the molar enthalpy of reaction of hydrogen at temperature T_0. The changes in the adiabatic flame temperature in Figure 4.26(a) were obtained by solving equation [4.91], by representing the molar heat capacities at a constant pressure using polynomials, whose coefficients are given by McBride et al. (1993). Without any excess air, the flame temperature is very high. This temperature decreases as the excess air increases.

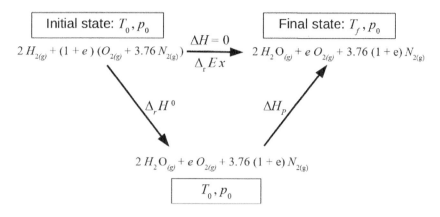

Figure 4.26. *Combustion of hydrogen using air as the oxidant*

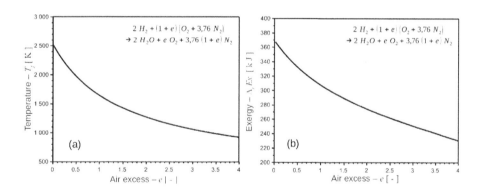

Figure 4.27. *Combustion of hydrogen using air as an oxidant: (a) adiabatic flame temperature as a function of the excess air and (b) exergy production as a function of the excess air*

The exergy Ex_0 of the fuel mixture with the initial temperature and pressure conditions can be determined via the method described above:

$$Ex_0 = R\,T_0 \left(2\,\ln(y_{0(H_2)}) + (1+e)\,\ln(y_{0(O_2)}) + 3.76\,(1+e)\,\ln(y_{0(O_2)})\right)$$

[4.97]

where the molar fractions of the mixture with the initial conditions are:

$$y_{0(H_2)} = \frac{2}{n_0} \qquad y_{0(O_2)} = \frac{1+e}{n_0} \qquad y_{0(N_2)} = \frac{3.76(1+e)}{n_0} \qquad [4.98]$$

with: $n_0 = 2 + 4.76(1+e)$

The exergy Ex_f of the combustion gases at the temperature T_f and the pressure p_0 can be calculated using equation [4.87], and the exergy production which results from the combustion is derived from this as follows:

$$\Delta_r Ex = Ex_f - Ex_0 \qquad [4.99]$$

Figure 4.27(b) shows the decrease in exergy production as a function of the excess air.

4.5.3. *Exergy of a fuel*

Between the initial state i and the final state f of a transformation of a closed system that exchanges the work W_e with the exterior, and with the heat quantity Q_e, then, according to the *first principle* of thermodynamics (equation [1.2]), and by neglecting any variations in the kinetic and potential energy at the moment, we obtain:

$$W_e = U_f - U_i - Q_e \qquad [4.100]$$

where U_f and U_i are the internal energies of the system in the final and initial states, respectively.

If the temperature T of the interface where the heat exchange takes place remains constant throughout the transformation, then, according to the *second principle* of thermodynamics (equations [1.11] and [1.12]), the entropy variation $DeltaS$ of the system can then be expressed as:

$$\Delta S = S_f - S_i = \frac{Q_e}{T} + \Delta S_{irr} \qquad [4.101]$$

S_{irr} is zero in the case of a reversible transformation, and strictly positive for a real transformation. By combining equations [4.100] and [4.101], we obtain:

$$W_e = U_f - U_i - T(S_f - S_i) + T \Delta S_{irr} \qquad [4.102]$$

In the case of a combustion reaction in which the energy released can be converted into mechanical energy, W_e is a negative quantity. The absolute value of W_e is maximal if $S_{irr} = 0$, i.e. *if the transformation is reversible*. The work exchanged $W_{e\,rev}$ during the reversible transformation is:

$$W_{e\,rev} = U_f - U_i - T(S_f - S_i) \qquad [4.103]$$

Introducing now the *useful work* W_m (equation [1.20]), we find:

$$W_{m\ rev} = U_f - U_i + p_0\,(V_f - V_i) - T\,(S_f - S_i) \qquad [4.104]$$

p_0 is the pressure of the environment that is chosen to be the reference, V_f and V_i are the volumes of the system in the final and initial states, respectively. Equation [4.99] could be generalized to take into account variations in the kinetic and situational potential (gravitational, field, etc.) energies of the system for the moment in question, by replacing U by $U + E_c + E_z$, where E_c and E_z are the kinetic and situational potential energies of the system.

Now, let us consider a *reactant mixture* that is initially under the ambient reference conditions (temperature T_0, pressure p_0), that evolves at a constant pressure and whose final product mixture of the reaction is brought back to the ambient conditions. The reversibility of the transformation forces the heat exchange interface temperature to be equal to T_0, and the expression [4.99] becomes:

$$\begin{aligned}W_m^0 &= U_f^0 - U_i^0 + p_0\,(V_f^0 - V_i^0) - T_0\,(S_f^0 - S_i^0) \\ &= H_f^0 - H_i^0 - T_0\,(S_f^0 - S_i^0) = G_f^0 - G_i^0 = \Delta G^0\end{aligned} \qquad [4.105]$$

Here, $U_{i,f}^0$, $V_{i,f}^0$, $S_{i,f}^0$, $H_{i,f}^0$, $G_{i,f}^0$ are the internal energies, volumes, entropies, enthalpies, and free enthalpies (Gibbs' function), respectively, of the system, whose conditions are the same as that of the environment in the initial and final states.

The *third principle* of thermodynamics, introduced by Walther Nernst (1844–1941) and Max Planck (1858–1947), can be stated as:

> The entropy of all pure crystalline bodies tends to zero, when the thermodynamic temperature tends to zero.

From the third principle of thermodynamics, it follows that the *standard entropy of a substance is a measurable quantity* (Figure 4.28), which can be obtained by measuring the amount of heat required to increase its temperature by a certain amount, via a reversible process.

Standard molar values (temperature $T_0 = 298$ K, pressure $p_0 = 0.1$ MPa) of enthalpies of formation, free enthalpies of formation and entropies of many species can be found in the literature. Table 4.6 shows some useful values for combustion.

For a chemical reaction that occurs at the reference pressure p_0:

$$\sum_i n_i\,A_i \longrightarrow \sum_j n'_j\,A'_i \qquad [4.106]$$

and whose reactants and products are at the reference temperature T_0, then the enthalpy of reaction $\Delta_r H^0$, the free enthalpy of reaction $\Delta_r G^0$ and the entropy change $\Delta_r S^0$ can be calculated using the following relations:

$$\Delta_r H^0 = \sum_j n'_j \, \Delta_f H^0_j - \sum_i n_i \, \Delta_f H^0_i \qquad [4.107]$$

$$\Delta_r G^0 = \sum_j n'_j \, \Delta_f G^0_j - \sum_i n_i \, \Delta_f G^0_i \qquad [4.108]$$

$$\Delta_r S^0 = \sum_j n'_j \, S^0_j - \sum_i n_i \, S^0_i \qquad [4.109]$$

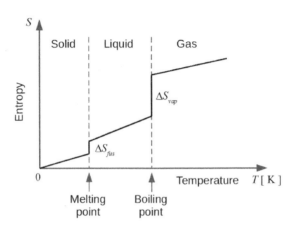

Figure 4.28. *Entropy as a function of the temperature*

Taking the combustion of methane with oxygen as an example:

$$CH_{4(g)} + 2\,O_{2(g)} \longrightarrow CO_{2(g)} + 2\,H_2O_{(g)} \qquad [4.110]$$

Using the values in Table 4.6:

$\Delta_r H^0 = (-393.5 + 2\,(-241.8)) - (-74.8 + 2\,(0))$

$\Delta_r H^0 = -802.3$ kJ/mol

which corresponds to the net molar heating value of methane.

$\Delta_r G^0 = (-394.4 + 2\,(-228.6)) - (-50.8 + 2\,(0))$

$\Delta_r G^0 = -800.8$ kJ/mol

Species	Formula	$\Delta_f H^0$ [kJ/mol]	$\Delta_f G^0$ [kJ/mol]	S^0 [J/(mol·K)]
Aluminum	$Al_{(s)}$	0	0	28.3
Nitrogen	$N_{2(g)}$	0	0	191.5
Carbon (diamond)	$C(s)$	0	0	2.37
Carbon (graphite)	$C(s)$	0	0	5.7
Hydrogen	$H_{2(g)}$	0	0	130.6
Oxygen	$O_{2(g)}$	0	0	205
Water	$H_2O_{(l)}$	-285.8	-237.2	69.9
	$H_2O_{(g)}$	-241.8	-228.6	188.7
Carbon monoxide	$CO_{(g)}$	-110.5	-137.3	197.9
Carbon dioxide	$CO_{2(g)}$	-393.5	-394.4	213.6
Nitrogen monoxide	$NO_{(g)}$	90.4	86.7	210.7
Nitrogen dioxide	$NO_{2(g)}$	33.8	51.8	211.2
Nitrogen peroxide	$N_2O_{2(g)}$	9.7	98.3	304.3
Methane	$CH_{4(g)}$	-74.8	-50.8	186.2
Ethane	$C_2H_{6(g)}$	-84.7	-32.9	229.5
n-Propane	$C_3H_{8(g)}$	-103.8	-23.5	269.9
n-Butane	$C_4H_{10(g)}$	-126.2	-17.1	310.1
n-Pentane	$C_5H_{12(l)}$	-173	9.2	262.7
	$C_5H_{12(g)}$	-146.4	-8.2	348.4
n-Hexane	$C_6H_{14(l)}$	-198.8	-3.8	296
	$C_6H_{14(g)}$	-167.2	0.2	396.8
n-Octane	$C_8H_{18(l)}$	-249.9	7.4	357.7
	$C_8H_{18(g)}$	-208.4	17.3	463.7
n-Dodecane	$C_{12}H_{26(l)}$	-351	28	493
	$C_{12}H_{26(g)}$	-291	50.2	623
n-Hexadecane	$C_{16}H_{34(l)}$	-456	58.5	586
Iso-octane	$C_8H_{18(l)}$	-260	5.8	219.4
Ethylene	$C_2H_{4(g)}$	52.3	68.1	329
Benzene	$C_6H_{6(l)}$	48.7	123	172.8
	$C_6H_{6(g)}$	82.9	129.7	269.2
Ethanol	$C_2H_5OH_{(l)}$	-277.6	-174.8	160.7
	$C_2H_5OH_{(g)}$	-235.3	-168.6	282
Ammonia	$NH_{3(g)}$	-46.2	-16.6	192.5

Table 4.6. Standard molar values of enthalpies of formation, free enthapies of formation and entropies of some species

As mentioned previously, $\Delta_r G^0$ corresponds to the *maximum useful work* for when the reactants and products of the reaction are in the standard state, which makes this quantity a possible choice for defining the fuel exergy.

$$\Delta_r S^0 = (213.6 + 2\,(188.7)) - (186.2 + 2\,(205))$$

$$\Delta_r S^0 = -5.2 \text{ J/(K} \cdot \text{mol)}$$

Using relation [4.105]:

$$\Delta_r G^0 = \Delta_r H^0 - T_0 \Delta_r S^0 = -802.3 \times 10^3 - 298\,(-5.2)$$

$$\Delta_r G^0 = -800.8 \text{ kJ/mol}$$

from which we find the value that was previously calculated using equation [4.108]. In the case of methane, we note the small difference between $\Delta_r H^0$ and $\Delta_r G^0$, which is also the case for many fuels, and justifies the fact that the lower heating value is often equated with the exergy of the fuel.

4.6. Conclusion

The reduction of gaseous and particulate pollutant emissions (soot), the thermokinetics of combustion mechanisms in the low-temperature range (< 1,000 K), which are responsible for the auto-ignition of fuels in thermal engines, are important subjects of research with regard to combustion. These topics require the understanding of particularly complex reaction flows due to their multi-physical and multi-scale nature.

The valorization of biomass, the use of new fuels which include oxygenated biofuels and new energy carriers as well as new modes of combustion (HCCI, etc.) all open up very wide fields of research.

4.7. References

Bibrzycki, J. and Poinsot, T. (2011). Examination of simplified mechanisms of CH4 combustion in N2/O2 and CO2/O2 atmosphere using mathematical modeling. *Archivum Combustions*, 31, 255–262.

Boivin, P., Jiménez, C., Sànchez, A.L., Williams, F.A. (2011). An explicit mechanism for H_2–air combustion. *Proceeding of the Combustion Institute*, 33(1), 517–523.

Borghi, R. and Destriau, M. (1995). *La combustion et les flammes*. Editions Technip, Paris.

Borgnakke, C. and Sonntag, R.E. (2012). *Fundamentals of Thermodynamics*, 8th edition. John Wiley & Sons, Hoboken.

Fenimore, C.P. (1971). Formation of nitric oxide in premixed hydrocarbon flames. *Symposium (International) on Combustion*, 13(1), 373–380.

Griffiths, J.F. and Barnard, J.A. (1995). *Flame and Combustion*, 3rd edition. Blakie Academic & Professionnal, Chapman & Hall, Glasgow.

Jones, W.P. and Linstedt, R.P. (1988). Global reaction schemes for hydrocarbon combustion. *Combustion and Flame*, 73, 233–249.

McBride, B.J., Gordon, S., Reno, M.A. (1993). Coefficients for calculating thermodynamic and transport properties of individual species. NASA Technical Memorandum 4513, October.

Miyamoto, N., Ogawa, H., Doi, K. (1990). Calculations of ignition lags for methane-air mixtures by chemical kinetics. *COMODIA 90 – Proceedings of the International Symposium on Diagnostics and Modeling of Combustion in Internal Combustion Engines*, Kyoto, 99–104.

Sawerysyn, J.-P. (1993). Les pouvoirs calorifiques. *Bulletin de l'union des physiciens*, 87(752), 401–411.

Titova, N.S., Torokhov, S.A., Starik, A.M. (2011). On kinetic mechanisms of n–decane oxidation. *Combustion, Explosion and Shock Waves*, 47(2), 129–146.

Zabetakis, M.G. (1965). Flammability characteristics of combustible gases and vapors. *U.S. Bureau of Mines, Bulletin*, 627.

Zeldovitch, Y.B., Sadovnikov, D.A., Kamenetskii, F. (1947). *Oxidation of Nitrogen in Combustion*. Institute of Chemical Physics, Moscow-Leningrad.

5

Engines with an External Heat Supply

Georges DESCOMBES[1] and Bernard DESMET[2]

[1]*CNAM, Paris, France*
[2]*INSA – HdF, Université Polytechnique Hauts-de-France, Valenciennes, France*

5.1. Introduction

In engines that use *external combustion*, also known as "*hot air engines*", no heat input to the working fluid is produced by combustion in the engine. Instead, the heat input is produced by external combustion, which may be continuous, and is achieved via a heat exchange through cylinder walls or heat exchangers.

Two types of external heat engines have been studied for many years:

– The *Stirling engine*, patented by Robert Stirling in 1816. It was realized predominantly by the Philips firm in the Netherlands during the 1830s.

– The *John Ericsson engine*, the first model of which dates back to 1833. This type of engine was produced between 1889 and 1900.

External heat engines are used for specific applications (military, space, solar energy), and they have the advantage of being able to operate using any form of energy source. Modern technologies also have allowed for improved performance. With the need for alternative energy sources and methods of energy recovery, these motors are now finding a renewed interest.

5.2. The Stirling engine

5.2.1. *Theoretical cycle*

Stirling engines are *externally heated, gaseous single-phase working fluid reciprocating heat engines*. They operate without any mass transfer to the outside and do not have any shut-off device for the gas circuit, in which the mass of the working fluid remains constant.

The *regenerator* allows the heat energy of the hot fluid flowing through it to be *stored*, and also *restores* this energy when the cold fluid flows through it. To allow for fairly rapid storage and retrieval operations, the material that accumulates the thermal energy can be, for example, a metallic foam. Figures 5.1 shows the different phases of the operating cycle of a Stirling engine.

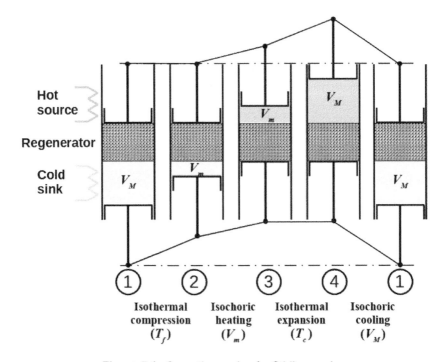

Figure 5.1. *Operating cycle of a Stirling engine*

Depending on the type of implementation, Stirling engines comprise one or more cylinders which constitute two cavities: the *hot cavity* receives the heat that is produced externally, and the *cold cavity* is in contact with the cold source. The regenerator is located between the hot and cold cavities.

The theoretical evolutions on pressure–volume $p - V$ and entropy $T - s$ diagrams are shown in Figure 5.2. The thermodynamic Stirling cycle is made up of two isothermal and two isochoric transformations (constant volume). The operating phases are made up of the following:

– 1–2: the *isothermal compression* to the temperature T_f of the low-temperature cavity. $|Q_{1\to2}|$ denotes the amount of heat given up by the working fluid to the cold source during the compression from the volume V_M to the volume V_m.

– 2–3: the *isochoric heating* (volume V_m) of the working fluid from the temperature T_c to the temperature T_h of the hot cavity. All or part of the heat is taken from the regenerator. In the case of partial recovery, $Q_{2\to3'}$ is the amount of heat returned by the regenerator, and $Q_{3'\to3}$ is the amount of heat supplied by the hot source.

– 3–4: the *isothermal expansion* to the temperature T_h of the high-temperature cavity. The necessary amount of heat $Q_{3\to4}$ is supplied by the hot source.

– 4–1: the *isochoric cooling* (volume V_M) from the temperature T_h to the temperature T_c. All or part of the heat extracted from the working fluid is stored by the regenerator. In the case of partial storage, only the part $|Q_{4\to1'}|$ is stored and the complement $|Q_{1'\to1}|$ is given to the cold source.

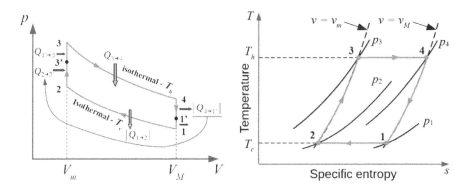

Figure 5.2. *Theoretical Stirling cycle*

For the study of the theoretical cycle, the simplifying assumptions are as follows:

– the working fluid is an *ideal gas* with the mass constant r and the heat capacity ratio $\gamma = c_p/c_v$;

– the hot and cold cavities are in *pressure equilibrium* at any given point in the cycle, which assumes that the pressure losses in the regenerator and the various connecting pipes are negligible.

The work exchanged with the fluid during the cycle is given by:

$$W_{cycle} = \oint_{cycle} -p\, dV = W_{1\to 2} + W_{3\to 4} \qquad [5.1]$$

By introducing the equation of state of perfect gases, the isothermal compression work $W_{1\to 2}$ at the temperature T_c can be written as:

$$W_{1\to 2} = \int_{(1\to 2)} -p\, dV = m\, r\, T_c \ln(\tau) \qquad [5.2]$$

where m is the mass of the evolving gas and $\tau = V_M/V_m$ is the compression volume ratio.

Since the internal energy of an ideal gas depends only on its temperature, its variation $\Delta U_{1\to 2} = 0$ and, according to the first principle of thermodynamics (equation [1.1]), the heat quantity exchanged with the cold source during the compression is:

$$Q_{1\to 2} = -W_{1\to 2} \qquad (Q_{1\to 2} < 0) \qquad [5.3]$$

Similarly, for isothermal expansion at the temperature T_h:

$$W_{3\to 4} = \int_{(3\to 4)} -p\, dV = -m\, r\, T_h \ln(\tau) \qquad [5.4]$$

$$Q_{3\to 4} = -W_{3\to 4} \qquad (Q_{3\to 4} > 0) \qquad [5.5]$$

The work involved during the cycle can be deduced from this as:

$$W_{cycle} = -m\, r\, (T_h - T_c) \ln(\tau) \qquad [5.6]$$

Theoretically, the amount of heat $|Q_{4\to 1}|$ given up by the working fluid during isochoric cooling 4–1 offsets that from the heat input $Q_{2\to 3}$ which is required for isochoric heating 2–3:

$$|Q_{4\to 1}| = Q_{2\to 3} = m\, c_v\, (T_h - T_c) \qquad [5.7]$$

where $c_v = r/(\gamma - 1)$ is the specific heat capacity at a constant volume.

In reality, the regenerator only allows for the recovery of the heat quantity $Q_{2\to 3'} < Q_{2\to 3}$, and the heat quantity $Q_{3'\to 3}$ will have to be taken from the hot source. Ultimately, the amount of heat Q_h supplied per cycle by the hot source is:

$$Q_h = Q_{3'\to 3} + Q_{3\to 4} \qquad [5.8]$$

The proportion of the thermal energy recovered can be characterized by introducing the regenerator *effectiveness* E_c:

$$E_c = \frac{Q_{2 \to 3'}}{Q_{2 \to 3}} = \frac{T_{3'} - T_c}{T_h - T_c} \quad [5.9]$$

$$Q_{3' \to 3} = m\, c_v\, (T_h - T_c)\, (1 - E_c) \quad [5.10]$$

The case with $E_c = 1$ (*perfect regenerator*) corresponds to the theoretical case in which all of the heat required for the isochoric heating 2–3 is recovered from the cooling 4–1, and $E_c = 0$ corresponds to the *no regenerator*.

By substituting expressions [5.5] and [5.10] into [5.8]:

$$Q_h = m\, c_v\, (T_h - T_c)\, (1 - E_c) + m\, r\, T_h\, \ln(\tau) \quad [5.11]$$

The thermal efficiency of the cycle can then be written as:

$$\eta_{th} = \frac{|W_{cycle}|}{Q_h} = \eta_c \frac{1}{1 + \dfrac{1 - E_c}{(\gamma - 1)\ln(\tau)} \eta_c} \quad \text{with: } \eta_c = 1 - \frac{T_c}{T_h} \quad [5.12]$$

In the case of a perfect regenerator ($E_c = 1$), the expression for the thermal efficiency reduces to the expression for the *Carnot efficiency* η_c. The curves in Figure 5.3 represent the evolution of the thermal efficiency for the Stirling cycle, with a fluid whose heat capacity ratio is $\gamma = 1.4$ as a function of the temperature ratio T_h/T_c for different values of the regenerator effectiveness E_c. The volumetric ratio is fixed at $\tau = 10$. A decrease in the regenerator effectiveness results in a decrease in thermal efficiency, and this decrease is the most noticeable in the vicinity of $E_c = 1$, which shows the necessity for a regenerator with high efficiency to obtain a yield similar to that of the Carnot efficiency.

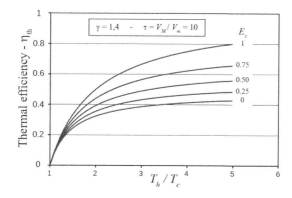

Figure 5.3. *Influence of the regenerator effectiveness*

In Figure 5.4, the efficiency curves as a function of the volumetric ratio τ are obtained for a fluid with a mass heat ratio of $\gamma = 1.4$ and a temperature ratio of $\frac{T_h}{T_c} = 4$.

The operation with a perfect regenerator results in a thermal efficiency equal to the Carnot efficiency, regardless of the volumetric ratio. In the case of an imperfect regenerator, the thermal efficiency decreases as the volumetric ratio decreases. However, if the efficiency of the regenerator remains fairly close to 1 (e.g. above 0.8), the decrease in efficiency remains low should the volumetric ratio be sufficiently large (e.g. approximately 6).

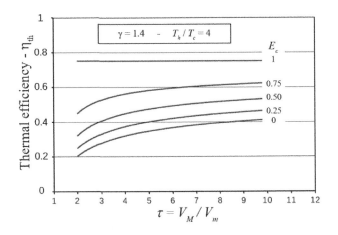

Figure 5.4. *Influence of the volume ratio*

5.2.2. *Characteristics of the Stirling engine*

Stirling engines are not widely used for waste heat recovery in vehicles, as they are not well suited for partial load operations (Bert 2012). Instead, they are used for space and military applications, as well as in micro-cogeneration applications. Thus, the Stirling cycle is widely used in cryogenic applications to produce low-temperature energy, and in a few rare stationary cogenerated heat engine applications.

Due to the *absence of valves* in Stirling engines, there is an *alternating transfer of fluid* between the hot and cold parts of the machine, which is also known as *oscillating flow*.

The Stirling engine comes in three different versions.

5.2.2.1. *Alpha version*

In the alpha version, a Stirling engine has two cylinders and two pistons.

Engines with an External Heat Supply 185

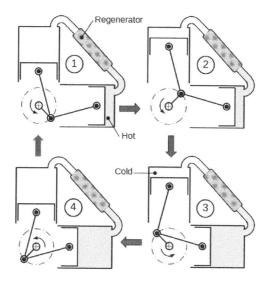

Figure 5.5. *Stirling engine: alpha version*

5.2.2.2. *Beta version*

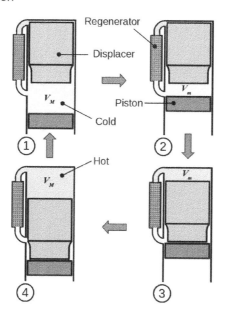

Figure 5.6. *Stirling engine: beta version*

The beta Stirling engine has a single cylinder in which the power piston and the displacer piston move. If pressure losses in the regenerator are neglected, then the hot and cold volumes are in a pressure equilibrium, and the displacer piston, whose job is to displace the working fluid from one cavity to the other, requires no work for it to be set into motion.

The theoretical displacements of the power and displacer pistons, shown in Figure 5.6, can be approximated as a rhomboidal crankcase (Figure 5.7).

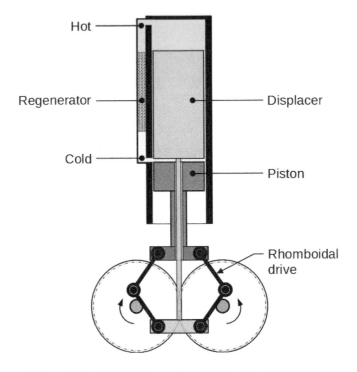

Figure 5.7. *Beta Stirling engine with the rhomboidal drive*

5.2.2.3. *Gamma version*

The gamma Stirling engines have two cylinders. In the first cylinder, the power piston moves, and in the second cylinder, the displacer piston moves.

The gamma version can be considered as a hybrid of the alpha and beta versions, and the phases of operation are those already described in section 5.2.1.

Engines with an External Heat Supply   187

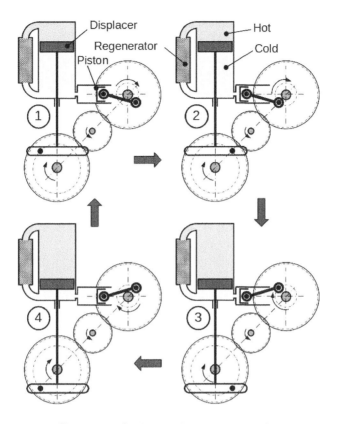

Figure 5.8. *Stirling engine: gamma version*

5.3. The Ericsson engine

5.3.1. *Operating principles*

An Ericsson engine (Figure 5.9) is an external heat input engine that generally operates in an open cycle, using a single-phase gaseous working fluid, namely air, in the case of an open cycle.

In its simplest form, an Ericsson engine comprises:

– a *positive displacement compressor*, in which the air is compressed from the ambient pressure p_a to the pressure p_e. T_c is the discharge temperature of the compressor;

– a *reheater exchanger*, which heats the previously compressed air using the heat extracted from the hot source. The pressure in the exchanger and the connecting pipes between the compressor and the expansion motor is assumed to be uniform and equal to p_e. The heat exchanger supplies the amount of heat Q_h necessary to raise the temperature of the air from T_c to T_h;

– a *positive displacement motor*, where the air expansion produces mechanical energy which is partly used to drive the compression machine, with the remainder left available to the user.

The compression and expansion cylinders are equipped with flaps or valves that isolate these cylinders, and, as a result, the exchanger does not contribute to any clearance volume.

The Ericsson engine can be compared with a gas turbine whose turbomachines are replaced by positive displacement machines, and whose combustion chamber is replaced by a heat exchanger that may use various sources of heat, for example solar thermal collectors (Alaphilippe et al. 2013) or boilers (Creyx 2014).

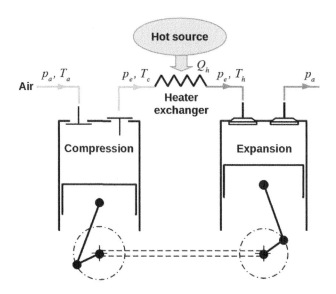

Figure 5.9. *Ericsson engine: operating principle*

5.3.2. *Theoretical cycles*

Figure 5.10 shows the evolution of the working fluid in the compressor and the expansion engine cycles of an Ericsson engine in the pressure–volume diagrams. The characteristics of the two machines are not necessarily identical. V_e, V'_e, ε, ε' are the volumes generated by the displacements of the pistons and the clearance volume coefficients, respectively, for the compressors and the expansion engines. Note also that the rotational speeds and therefore the number of cycles described may be different for the two machines.

The cycle of the *compressor machine* includes the following phases:

– 1–2: *compression*, assumed adiabatic and reversible from the ambient pressure p_a to the pressure of p_e of the exchanger;

– 2–3: *backflow* at the constant pressure p_e of the exchanger and at the constant temperature ($T_2 = T_3 = T_c$). Point 2 corresponds to the opening of the discharge valve, which is assumed to be instantaneous as well as taking place under a negligible pressure difference between the cylinder and the discharge line;

– 3–4: *expansion* of the gases contained within the clearance volume from the pressure p_e to the ambient pressure p_a, assumed adiabatic and reversible. Point 4 corresponds to the opening of the intake valve with the same hypothesis as for the discharge valve;

– 4–1: *intake* under the ambient conditions (pressure p_a, temperature T_a).

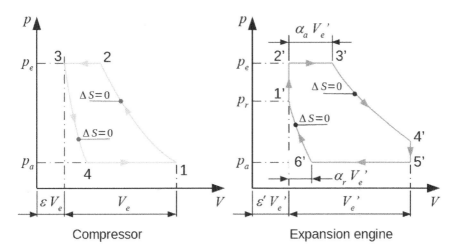

Figure 5.10. *Theoretical compressor and expansion motor cycles*

Air is assimilated into an ideal mass constant r and, according to the law of reversible adiabatic:

$$\frac{T_2}{T_1} = \frac{T_3}{T_4} = \frac{T_c}{T_a} = \left(\frac{p_e}{p_a}\right)^{\frac{\gamma-1}{\gamma}} \qquad [5.13]$$

The air masses contained in the cylinder at the end of the intake and at the end of discharge are, respectively:

$$m_1 = \frac{p_a\,(1+\varepsilon)\,V_e}{r\,T_a} \qquad m_3 = \frac{p_e\,\varepsilon\,V_e}{r\,T_c} \qquad [5.14]$$

Assuming the reversible adiabatic law, the mass m_c of air transferred per cycle from intake to discharge can be easily expressed as:

$$m_c = m_1 - m_3 = \frac{p_a V_e}{r T_a} \left(1 + \varepsilon - \varepsilon \left(\frac{p_e}{p_a}\right)^{\frac{1}{\gamma}}\right) \quad [5.15]$$

The transferred mass m_c decreases as the compression volume ratio p_c/p_a increases, canceling when:

$$\frac{p_e}{p_a} = \left(\frac{1+\varepsilon}{\varepsilon}\right)^{\gamma} \quad [5.16]$$

Using the first principle of thermodynamics (equation [1.2]), the work W_c exchanged per cycle with air is given by:

$$W_c = \oint_{cycle} -p\,dV = W_{1\to 2} + W_{2\to 3} + W_{3\to 4} + W_{4\to 1}$$
$$= \Delta U_{1\to 2} - p_e (V_3 - V_2) + \Delta U_{3\to 4} - p_a (V_1 - V_4) \quad [5.17]$$

and, taking into account the properties of an ideal gas ($u = c_v T$; $c_v = r/(\gamma - 1)$):

$$W_c = m_c c_p (T_c - T_a) \quad [5.18]$$

The work done per unit mass of air transferred is therefore:

$$w_c = \frac{W_c}{m_c} = c_p (T_c - T_a) = \frac{\gamma r T_a}{\gamma - 1} \left(\left(\frac{p_c}{p_a}\right)^{\frac{\gamma-1}{\gamma}} - 1\right) \quad [5.19]$$

which does not depend on the clearance volume.

The cycle of the *expansion motor* has the following phases:

– 1'–2': *filling of the clearance volume* (clearance volume $\varepsilon' V'_e$; V'_e: volume displaced by the piston, ε': clearance volume coefficient). At the opening of the inlet valve (point 1'), which is assumed to be instantaneous, the air delivered by the exchanger at the pressure p_e and the temperature T_h fills the clearance volume until it reaches the pressure p_e. We assume that this evolution is adiabatic but, as the flow losses at the valve passage are non-negligible, it is irreversible.

– 2'–3': *intake* of the air volume $\alpha_a V'_e$ (α_a: admission coefficient) at the constant pressure p_e and temperature T_h. Point 3' corresponds to the inlet valve closing.

– 3'–4': adiabatic and reversible *expansion* of the working fluid until the exhaust valve opens.

– 4'–5': *assumed instantaneous expansion* from the pressure $p_{4'}$ to the ambient pressure p_a. For the gases that remain in the cylinder, it is assumed that this expansion is adiabatic and reversible.

– 5'–6': *discharge* of air at the constant pressure p_a and at a constant temperature until the exhaust valve closes (point 6').

– 6'–1': adiabatic and reversible *recompression* of the gases contained within the cylinder up to the pressure p_r and the temperature T_r, described by the coefficient α_r. Touré (2010) shows that this recompression improves the performance of the Ericsson engine.

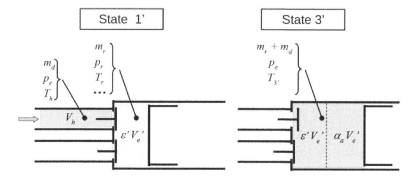

Figure 5.11. *Expansion motor: air intake*

To determine the mass of air taken in per cycle in the expansion engine, the intake phase is studied between the instant of the intake valve opening (state 1') and the instant of its closing (state 3'). In state 1' (Figure 5.11), the thermodynamic system under consideration consists of the mass m_r of air contained within the clearance volume (volume: $\varepsilon' \, V'_e$) at the pressure p_r and the temperature T_r, and the mass m_d admitted per cycle by the engine at the pressure p_e of the reheater exchanger, and at the temperature T_h occupying the volume V_h in the intake circuit. At the end of the intake (state 3'), the isolated system occupies the volume $(\varepsilon' + \alpha_a) \, V'_e$ of the cylinder at the pressure p_e and the temperature $T_{3'}$.

Assuming that any heat exchanges are negligible, according to the first law of thermodynamics:

$$W_{e-(1' \to 3')} = \Delta U_{1' \to 3'} \qquad [5.20]$$

with:

$$\begin{aligned} W_{e-(1' \to 3')} &= p_e \, V_h - p_e \, \alpha_a \, V'_e \\ &= m_d \, r \, T_h - p_e \, \alpha_a \, V'_e \end{aligned} \qquad [5.21]$$

$$\Delta U_{1'\to 3'} = (m_r + m_d)\, c_v\, T_{3'} - (m_r\, c_v\, T_r + m_d\, c_v\, T_h) \qquad [5.22]$$

By substituting expressions [5.21] and [5.22] into equation [5.20], and by taking into account the ideal gas properties of air, the mass m_d taken in per cycle can be put in the form:

$$m_d = \frac{p_e\, V'_e}{c_p\, T_h} \left[\alpha_a + \frac{\varepsilon' + \alpha_a}{\gamma - 1} - \frac{p_r}{p_e}\frac{\varepsilon'}{\gamma - 1} \right] \qquad [5.23]$$

and:

$$T_{3'} = \frac{p_e\,(\varepsilon' + \alpha_a\, V'_e)}{(m_r + m_d)\, r} \quad \text{with: } m_r = \frac{p_r\, \varepsilon'\, V'_e}{r\, T_r} \qquad [5.24]$$

The conservation of the air mass flow between the compressor and the expansion motor must be respected. If n' represents the number of expansion motor cycles per compressor cycle, then the mass conservation equation can be written as:

$$m_c = n'\, m_d \qquad [5.25]$$

where m_c and m_d are given by relations [5.15] and [5.24].

Having defined the temperature $T_{3'}$, the other temperatures of the expansion motor cycle can be easily deduced:

$$T_{4'} = T_{3'} \left(\frac{\varepsilon' + \alpha_a}{\varepsilon' + 1} \right)^{\gamma - 1}$$

$$T_{5'} = T_{3'} \left(\frac{p_a}{p_e} \right)^{\frac{\gamma-1}{\gamma}} = T_{6'} \qquad T_{1'} = T_{3'} \left(\frac{p_r}{p_e} \right)^{\frac{\gamma-1}{\gamma}} \qquad [5.26]$$

and the pressure p_r is deduced from the recompression coefficient α_r by:

$$p_r = p_a \left(\frac{\varepsilon' + \alpha_r}{\varepsilon'} \right)^{\gamma} \qquad [5.27]$$

The exhaust temperature of the expansion engine can be determined as in the case of an internal combustion engine (see section 2.5.3).

The work W'_{cycle} exchanged per cycle with the air in the expansion engine is obtained from:

$$W'_{cycle} = \oint_{cycle} -p\, dV = W_{2'\to 3'} + W_{3'\to 4'} + W_{5'\to 6'} + W_{6'\to 1'}$$
$$= -p_e(V_{3'} - V_{2'}) + \Delta U_{3'\to 4'} - p_a(V_{6'} - V_{5'}) + \Delta U_{6'\to 1'} \qquad [5.28]$$

$$W'_{cycle} = -p_e\, \alpha_a\, V'_e + (m_r + m_d)\, c_v\, (T_{4'} - T_{3'})$$
$$- p_a\, (1 - \alpha_r)\, V'_e + m_r\, c_v\, (T_r - T_{6'}) \quad [5.29]$$

The theoretical thermodynamic efficiency compares the useful work to the quantity of heat supplied by the exchanger:

$$\eta_{th} = \frac{n'\, |W'_{engine\ cycle}| - W_{compressor\ cycle}}{m_c\, c_p\, (T_h - T_c)} \quad [5.30]$$

5.3.3. Improvements of the Ericsson engine

5.3.3.1. Exhaust heat recovery of the expansion engine

When the temperature at the expansion motor exhaust is sufficient, a *recovery exchanger* placed between the compressor discharge and the heater exchanger (Figure 5.12) ensures the preheating of the working fluid. The reduction in the amount of heat provided by the hot source obviously results in an improvement in the thermodynamic efficiency of the Ericsson engine.

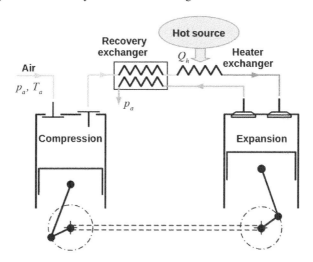

Figure 5.12. *Exhaust heat recovery of the expansion engine*

5.3.3.2. The Ericsson engine in a closed configuration

The use of a working fluid other than air, or a compressor intake pressure other than the atmospheric pressure, requires a *closed configuration* for the circuit of the working fluid (Figure 5.13).

In a closed configuration, the Ericsson engine must include a *cold heat exchanger*, or a *cooler*, to restore the heat quantity Q_c to the cold sink.

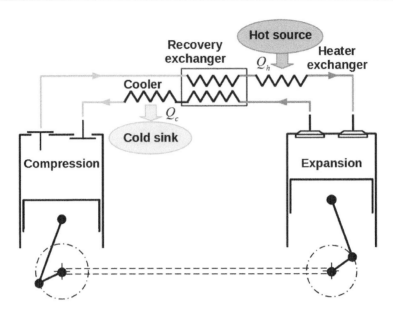

Figure 5.13. *The Ericsson engine in a closed configuration*

5.4. Perspectives

5.4.1. *Advantages and disadvantages of Stirling and Ericsson engines*

Unlike the reciprocating internal combustion engine, or to a lesser extent the gas turbine, these engines can operate with *any form of heat input*, as this heat comes from an external source. Traditional fossil fuels, as well as renewable energies such as biomass and solar energy, can be used.

Due to the *absence of valves*, during the expansion in the Stirling engine (Stouffs 2009), it is the fluid in the entire machine that is expanded. The exchangers therefore represent a large *clearance volume*, and their size must be limited in contradiction to the fact that efficient heat transfer requires a large exchanger. This problem does not exist in the Ericsson engine (Stouffs 2009), due to the *presence of valves*. On the contrary, because of its distribution components, it presents greater mechanical losses, dropping of pressure at the valves, increased noise, reduced reliability and greater complexity than the Stirling engine.

In the Stirling engine, we note that at the end of the compression, the fluid transferred to the hot side first passes through the cooler before being reheated in the regenerator and the heater. A similar problem appears during the transfer phase to the cold side. This problem is naturally avoided in the Ericsson motor because of the separate circuits.

These external combustion engines are preferred for stationary operations since their thermal inertia and response times do not lend themselves well to fluctuating power demands in an unsteady operation.

5.4.2. *Perspectives of evolution of external combustion machines in the new decarbonized energy landscape*

New perspectives such as the thermodynamic conversion of solar energy or the recovery of waste are being addressed by Stirling and Ericsson engines with an external heat supply.

A *micro-cogeneration* system, based on an Ericsson engine, coupled to a natural gas combustion system produces electricity and heat.

The necessary decarbonization of buildings and heating can benefit greatly from these external heat input machines in their specialized field as stationary applications. In the field of hybridization and polygeneration, their ability to produce additional mechanical energy at maximum efficiency with various hot sources of carbon-free heat is noteworthy. In this sense, Stirling and Ericsson derivative cycles are of notable interest. It is a prospective avenue for further development in the analysis of life-cycle resources, exergy analysis and the economic cycle of value.

Applications that use *solar energy* can also be mentioned, which use mirrors along with a focusing system to concentrate the sun's rays onto a receiver that converts light energy into thermal energy, so that a Stirling engine can produce electricity and heat. Harnessing thermal energy to heat buildings with combined heat and power (CHP) systems is an efficient way to supply energy locally, in addition to limiting efficiency losses that come from energy transportation.

Another alternative solution to meet the energy needs of individual households consists of coupling a micro-CHP boiler with a Stirling engine. In France and still in its infancy, micro-cogeneration technology uses the heat emitted by thermal equipment to produce electricity. In addition to hot water used for heating and sanitary needs, a micro-cogeneration boiler provides the electrical supply for an individual home. The operating principle consists of coupling a burner with a Stirling engine, transforming thermal energy into mechanical energy. A fluid (helium, for example) is subjected to a cycle comprising the following four phases: heating, expansion, cooling and compression.

5.5. References

Alaphilippe, M., Perier-Muzet, M., Sène, P., Stouffs, P. (2013). Étude d'un moteur Ericsson couplé avec un concentrateur solaire cylindro-parabolique. *Congrès français de thermique*, SFT, Gérardmer, 28–31 May 2013.

Bert, J. (2012). Contribution à l'étude de la valorisation des rejets thermiques – Étude et modélisation de moteurs Stirling. Thesis, Université de Bourgogne, 26 November 2012.

Creyx, M. (2014). Étude théorique et expérimentale d'une unité de micro-cogénération biomasse avec moteur Ericsson. Thesis, Université de Valenciennes et du Hainaut-Cambrésis (UVHC), 14 November 2014.

Stouffs, P. (2009). Les moteurs à apport de chaleur externe. *10ème Cycle de conférences CNAM/SIA*, Paris, March 2009.

Touré, A. (2010). Étude théorique et expérimentale d'un moteur Ericsson à cycle de Joule pour conversion thermodynamique de l'énergie solaire ou pour micro-cogénération. Thesis, Université de Pau et des Pays de l'Adour, 18 November 2010.

6
Energy Recovery – Waste Heat Recovery

Mohamed MEBARKIA
Larbi Tebessi University, Algeria

6.1. Waste energy recovery

6.1.1. *Energy balance of an internal combustion engine*

The sources of energy losses in an internal combustion engine can be found in the exhaust circuit, cooling circuits (water, oil, supercharged air if the engine is supercharged) and by the transfer of heat via convection and radiation.

If we assume that the mass of fuel in the cylinder is negligible compared to the mass of air, and assume that the air and exhaust gases are ideal and have the same heat capacities, then, according to the first law of the thermodynamics applied to the entire internal combustion engine and its supercharging system (Figure 6.1), we have in the established regime:

$$P_{mech} = \dot{m}_f\, LHV - \dot{m}_a\, c_{pa}\, (T_{exh} - T_{asp}) - \dot{m}_e\, c_{pe}\, \Delta T_{coolant} \\ - \dot{m}_l\, c_{pl}\, \Delta T_{oil} - \Phi_{charge\ air} - \Phi_{conv-rad} \quad [6.1]$$

where P_{mech} is the mechanical power produced, $\dot{m}_f\, LHV$ is the thermal power provided by the fuel ($\dot{m}_f$: mass flow rate of the fuel, LHV: lower heating value

Thermodynamics of Heat Engines,
coordinated by Bernard DESMET.
© ISTE Ltd 2022.

of the fuel), $\dot{m}_a\, c_{pa}\, (T_{exh} - T_{asp})$ is the thermal power lost at the exhaust ($\dot{m}_a$: mass flow rate of the air, c_{pa}: specific heat capacity at a constant air pressure, T_{exh}, T_{asp}: exhaust and intake temperatures), $\dot{m}_e\, c_{pe}\, \Delta T_{water}$ is the thermal power lost by the coolant ($\dot{m}_e$: mass flow rate, c_{pe}: specific heat capacity, $\Delta T_{coolant}$: increase in the coolant temperature), $\dot{m}_l\, c_{pl}\Delta T_{oil}$ is the thermal power lost by the oil ($\dot{m}_l$: mass flow rate, c_{pl}: specific heat capacity, ΔT_{oil}: increase in the oil temperature), $\Phi_{charge\ air}$ is the heat flux corresponding to the cooling of the charged air and $\Phi_{conv-rad}$ is the heat flux lost by convection and radiation at the walls of the engine and its charge system.

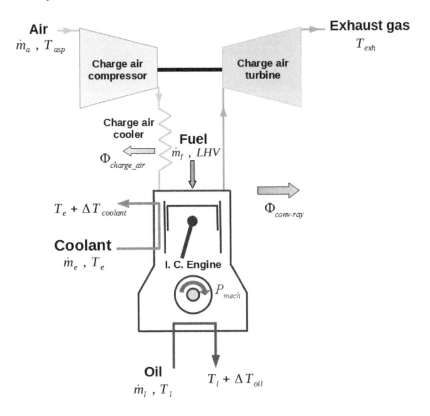

Figure 6.1. *Energy balance of an internal combustion engine*

The relevance of the quantity of thermal energy which dissipates to the exterior by convection–radiation is a debatable, and hence it is a delicate point to include in the energy balance. This is because in general, this term is not obtained from the heat transfer relations, but rather is deduced by looping the energy balance terms. Since the other terms in the balance accumulate errors, the looping of the balance may be an approximation, and the estimation of the losses by convection–radiation is thus questionable.

Of course, the energy balance of an engine depends on its field of application (Figure 6.2), meaning that the balance of a large industrial engine will have different values from that of an automobile engine, an agricultural engine or an engine with a smaller displacement, etc. The amount of lost energy for industrial engines is about 50–60%, and is higher for automotive engines (60–80%).

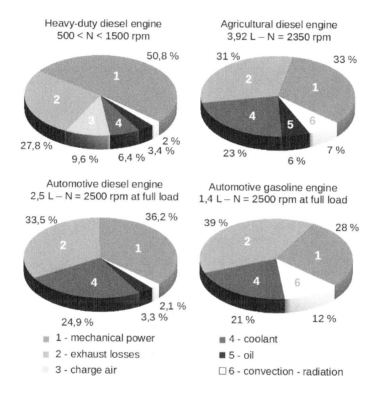

Figure 6.2. *Comparison of the heat balances for four diesel engines, depending on their fields of application from Descombes (2003)*

The heat flow in the cooling circuits of an industrial engine represents about 20% of the total power introduced, half of which comes from the charged air. For smaller engines, the heat flow in the cooling circuits is higher and depends on the application of the engine: the coupling of aerodynamic phenomena in an engine is particularly acute in the current generation of direct-injection engines, especially in those with reduced displacement or are greatly supercharged.

6.1.2. Degradation of mechanizable energy into uncompensated heat

6.1.2.1. Ideal reversible cycle

The concept of *mechanizable energy* can be established by considering a piston engine, for which the engine fluid at its terminals is cyclically extracted from the inlet at the pressure p_1 and discharged into the outlet at the pressure p_2 (Figure 6.3). Here, $E_n = E_n(T)$ is the thermal energy that is isothermically extracted from emission source of a *dithermal reversible* machine. Applying the *first law* of thermodynamics with respect to an open system leads to the following expressions:

$$[ex + an]_1^2 = u_2 + p_2 v_2 + \frac{V_2^2}{2} - u_1 - p_1 v_1 - \frac{V_1^2}{2} \qquad [6.2]$$

$$[ex + an]_1^2 = h_2 + \frac{V_2^2}{2} - h_1 - \frac{V_1^2}{2} \qquad [6.3]$$

where ex is the exergy that represents the mechanizable work produced on the crankshaft in the established cyclic operating regime, an is the non-mechanizable heat energy lost to the external medium (the anergy), u is the specific internal energy, V is the velocity of a fluid particle, $h + V_2^2/2$ is the total enthalpy of the fluid particle, p is the pressure and v is the specific volume of the elastic fluid which takes part in the cyclic transformation.

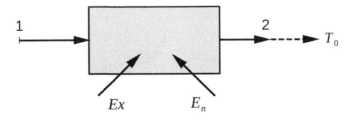

Figure 6.3. *Conversion of energy in a heat engine*

The *second law* of thermodynamics, when applied to thermal irreversibilities in the terminology of Emile Jouguet, determines the quality of the conversion processes that transforms chemical energy into work and uncompensated heat. It gives rise to the following relation:

$$T\, ds = dq + df = d\,an \qquad [6.4]$$

where $T\, ds$ is the elementary entropy variation at the temperature T, dq is the quantity of elementary heat exchanged with the exterior, df is the term for the dissipated mechanizable energy and $d\,an$ is the non-mechanizable energy.

6.1.2.2. Actual irreversible cycle

The concept of mechanizable energy allows us to evaluate the maximal work which is most likely to be obtained theoretically from a reversible engine, whose mixture of air–fuel reactant taken in under ambient conditions would evolve first to the post-combustion state, before returning themselves to the original state with ambient conditions.

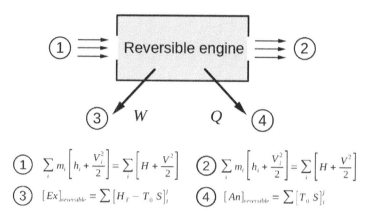

Figure 6.4. *Diagram representing the generation of energy according to an ideal reversible cycle*

The irreversibilities inherent to the actual process of converting heat energy into mechanical energy clearly and significantly reduce the efficiency of the transformation in the piston engine. The sources of irreversibility have mechanical and thermal origins, but may also have chemical, aerodynamic or diffusivity origins. In the context of the first law of thermodynamics, they give rise to a systematic reduction in the mechanical work $w_{irreversible}$ recovered from the crankshaft, and an equivalent increase in the amount of heat $q_{irreversible}$ returned to the atmosphere, according to the relation:

$$\Delta_1^2 [w + q]_{reversible} = \Delta_1^2 [w + q]_{irreversible} \qquad [6.5]$$

Any actual cycle has the effect of reducing the mechanical work generated from the transformation, as well as equivalently increasing the quantity of heat returned to the receiving sink. The heat is then ultimately transformed back into work with an efficiency lower than the Carnot efficiency and, limited by thermal irreversibilities, the second law of thermodynamics can be expressed in the form:

$$ds = \sum_j \frac{dq_j}{T_j} + \sum_k \frac{df_k}{T_k} = \sum_p \left[\frac{d\,an}{T}\right]_p \qquad [6.6]$$

Combining the energy [6.5] and the entropy balances [6.6], we can rephrase the concept of mechanizable energy ex picked up by the motor shaft for the actual operating conditions of the engine, according to the relation for the open system as:

$$ex = \left[h + \frac{V^2}{2} - \sum_j T_j\, s_j\right]_{in} - \left[h + \frac{V^2}{2} - \sum_j T_j\, s_j\right]_{out} \qquad [6.7]$$

where the subscripts in and out identify the input and output terminals of the heat engine, respectively.

The relation:

$$ex = [u - T_0\, s]_1 - [u - T_0\, s]_2 \qquad [6.8]$$

re-expresses the conversion of energy restricted to terminals 1 and 2 in the closed system determined by the open interval (RFA, AOE). Figure 6.5 illustrates the control volume of such an engine, where the chemical energy E_n can be decomposed into the exergy Ex picked up by the crankshaft, and the anergy An that makes up the rejection term, for which a partial recovery can be performed to increase the primary energy usage rate for the engine.

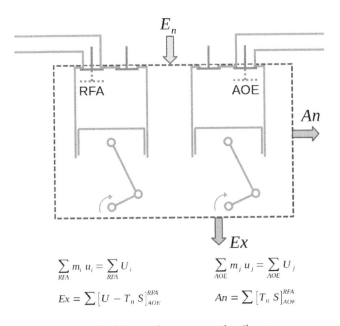

Figure 6.5. Schematics representing the energy generation from an actual cycle

Consequences of this degradation of mechanizable energy into uncompensated heat has been studied in section 6.1.1 in terms of heat balances. Hence, it is necessary whenever possible to give priority to the recovery of waste energy, if we are to maximize the efficiency when transforming primary energy. It is important to enhance the synergy between turbomachinery and alternative internal combustion engines, such as external combustion engines (Pluviose 2009). Another important application is the simultaneous production of electrical and thermal energies in the form of heat and refrigeration.

6.1.3. *Exergy balance in internal combustion engines*

As seen in section 6.1.1, internal combustion engines transform the chemical energy in a fuel into mechanical power with a limited chemical-to-mechanical energy conversion efficiency of less than 40% for automotive applications, and about 50% for stationary and marine applications. This means that in the most commonly used operating applications, more than 60% of this chemical energy in the fuel is lost as uncompensated heat. A fraction of this energy is rejected in the exhaust gases, some is lost in the water and oil cooling systems, and the rest is lost in the auxiliaries. For this purpose, the energy balance for the whole operational range should be analyzed and is usually represented in the form of heat balances, also called Sankey diagrams.

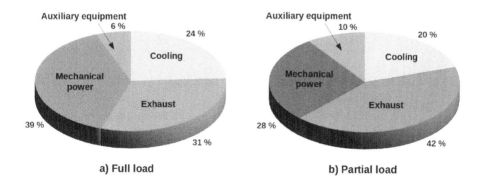

Figure 6.6. *Energy distributions for a turbo-supercharged diesel engine at full and partial loads*

The result, obtained from a numerical study of an agricultural tractor engine, is presented in Figure 6.6 (Milkov et al. 2014). Modern diesel engines operate with a maximum effectiveness at a full load whose rotational velocity is in a vicinity close to the maximum engine torque, and where the energy dissipated in the exhaust gases is approximately equal to that lost in the cooling system. For a partial load, engine effectiveness is lower and the exhaust energy increases significantly, while mechanical losses in the auxiliary equipment vary with the engine speed.

This energy balance allows us to estimate the engine efficiency in terms of the first law of thermodynamics, as well as the quantity of heat rejected in the exhaust gases and the cooling system. However, this balance cannot reveal the recovery potential of this lost energy. For this reason, an *exergy analysis* which combines the first two laws of thermodynamics is necessary to evaluate the *recovery potential* in the exhaust gases and in the cooling system.

The exergy of a thermodynamic system represents the work that can be obtained when the thermodynamic system has reached its equilibrium point. In other words, it is the energy available that is in the system. Provided that material transport is neglected, the exergy of the exhaust gases and the cooling system is estimated by the relation:

$$ex = h - h_0 - T_0 (s - s_0)$$ [6.9]

where h and s are the enthalpy and entropy of the *heat source*, and h_0, s_0 and T_0 are *reference parameters*, respectively. The reference temperature T_0 is considered in this case to be equal to 293 K. The exergy, which is calculated as a percentage of the lower heating value (LHV) for the fuel in the exhaust and cooling system, is presented in Figure 6.7.

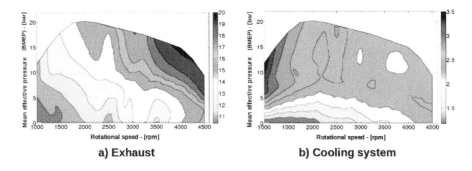

a) Exhaust   b) Cooling system

Figure 6.7. *Exergy of a) the exhaust gases and b) the cooling system for a diesel engine, expressed as a percentage of the LHV for the fuel*

Based on an exergy analysis of the exhaust gases, the overall engine efficiency can be improved by 10–20% and the maximum potential for recovery is obtained when operating at nearly the maximal engine power (Abusoglu and Kanoglu 2009). The exergy analysis presented in Figure 6.7 also shows that a heat recovery system can increase the overall engine efficiency by a few percent, depending on the engine speed and fuel load (Punov et al. 1993). Unsurprisingly, the recovery potential of the cooling system is less than that of the exhaust gases, due to the moderate temperature levels of the water and oil coolants.

6.1.4. *Concept of energy recovery*

The concept of energy recovery takes two distinct and complementary forms: one concerns the *exergy point of view*, that intends to maximize the mechanical energy produced in terms of the torque and power, along with the aim of reducing consumption and pollutants by as much as possible. The second point of view concerns itself with *optimizing the recovery of heat from emissions* in terms of the cooling and exhaust circuits of the engine. These emissions can be exploited to produce additional *mechanical or electrical energy* on one hand, or *thermal energy* in the form of heating or cooling, on the other.

Non-mechanized energy, namely discharged heat, can reach 60–80% of the energy introduced depending on the operational state of the engine. Since most of the chemical energy in the fuel is not converted into useful work, it is essential to consider solutions that will recover this part of the energy, which would then increase the overall efficiency of the transformation by recovering some of this discharged thermal energy. This recovery occurs primarily in the cooling and exhaust circuits.

With the recovered energy, it is possible to generate heat (hot water, steam, hot air, etc.), cold, mechanical energy (electricity production, supercharging via a turbocharger) and, if the exhaust gases are oxygen-rich, they can serve as the oxidizer in the post-combustion phase. Partial recovery of wasted heat energy allows for a significant reduction in primary energy consumption and emissions.

Cogeneration, which means the combined production of mechanical and thermal energy (used especially in the production of electrical energy) from the same energy source, makes it possible to fully exploit the fuel and therefore reduce carbon dioxide emissions (CO_2).

For an urban driving cycle of an automobile, the steady-state conditions are not reached since the engine operates only at a partial load and its efficiency is quite low. For example, for a standardized European cycle (ECE 15) with an average operating state of 2,000 rpm and 10 m · N, the overall efficiency is 9% (Trapy 1981).

In the exhaust circuit, the temperature and the flow rate of the gases are often low, and therefore the enthalpy contained in the gases is itself low too, exacerbating any possible recovery attempts from the circuit. On the other hand, it is almost always possible to find a solution for the cooling circuits.

6.2. Cogeneration in industrial facilities

6.2.1. *Cogenerating gas turbines*

For a gas turbine, the main source of discharged heat is concentrated in the exhaust. Figure 6.8 illustrates the case of an installation with a gas turbine cogenerator in its standard mode of operation. The mechanical power supplied to the alternator is 1,050 kW, the mass flow rate of the exhaust gases is 5.45 kg/s and the thermal power recovered is equal to 2,175 kW obtained by cooling the exhaust gases from 500°C to 150°C. The mass flow rate of hot water is 95,000 kg/h, while the saturated steam is produced at approximately 33,000 kg/h with a pressure of 4 bar. The conversion of primary energy into mechanical and thermal energy has an efficiency of about 85%.

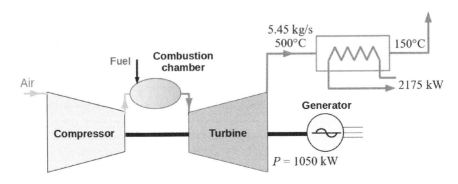

Figure 6.8. *Energy recovery with a gas turbine from Boudigues et al. (2001)*

Figure 6.9 illustrates the *preferred fields of use* for internal combustion engines, as well as for gas and steam turbines, and represents their overall efficiency as a function of the power production output per unit.

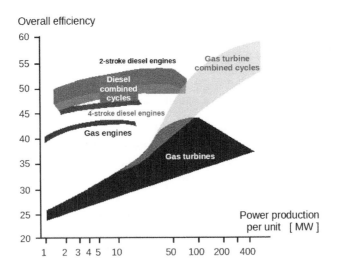

Figure 6.9. *Classification of cogenerating heat engines*

For powers greater than a few tens of MW, the preferred technologies are those with gas turbines and combined cycles, in which a turbo-alternator is connected to a gas turbine. The range of electrical power evolves from a few tens to a few hundreds of MW, and turbo-alternator technologies are divided between back-pressure and extraction cycles (Pluviose 2009). For low to intermediate powers in the range from a few tens of kW up to 10 MW, diesel engines are preferred for which the intrinsic efficiency is greater than that of turbomachines.

Figures 6.10 and 6.11 schematically represent an exhaust energy recovery system that uses a cogenerating gas turbine and a combined cycle, respectively.

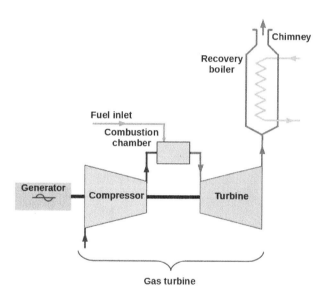

Figure 6.10. *Schematics of a cogenerating simple-cycle gas turbine*

6.2.2. *Cogenerating diesel engine*

Figure 6.12 (Descombes et al. 1999) shows the diagram case of an industrial turbocharged diesel engine with cogeneration, to produce electrical energy and hot water to an urban city.

The two 18-cylinder engines are powered using heavy fuel oil, with a viscosity of 180 centistokes at 50°C and a sulfur content less than 1.5%. The electrical power generated by the set of two alternators is 38.5 MWe.

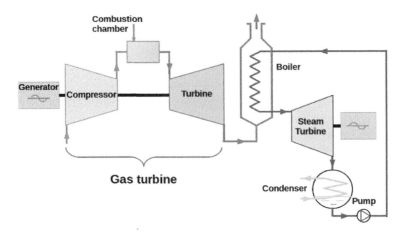

Figure 6.11. *Schematics of a gas and steam turbine arranged in a combined cycle*

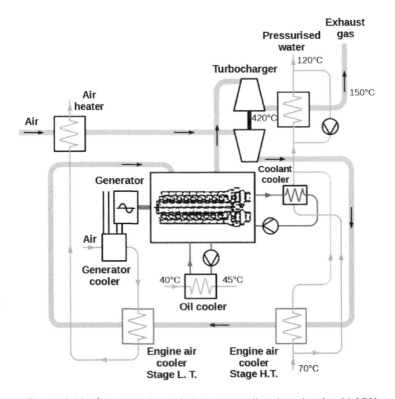

Figure 6.12. *Cogeneration unit that uses a diesel engine 2 x 20 MW*

The cogeneration process (Figure 6.13) allows us to obtain a simultaneous thermal power equal to 30 MW of hot pressurized water at 120°C, and 1 MW of low temperature water at 70°C. The overall efficiency of this power plant unit is approximately 80%, and the annual fuel saving compared with a non-cogeneration application is about 10,000 TOE.

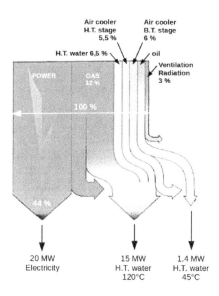

Figure 6.13. *Waste heat recovery within a diesel engine from Boudigues et al. (2001)*

6.2.3. *Comparative cogeneration efficiencies*

In the two preceding examples, the conversion of fuel into primary energy has an efficiency of about 80%. Nevertheless, it is observed that the ratio between the mechanical energy produced on the shaft of the diesel engine with the recoverable heat energy is reduced, when compared to that of a gas turbine.

In fact, the use of energy recovery at the terminals of a diesel engine must accept both the energy temperature level and the thermal-to-mechanical power ratio that is imposed by the engine. However, these two values can be very different from one industry to another, and they can vary between seasons or even throughout the day.

A major advantage of the gas turbine over the reciprocating engine lies in its ability to adapt to various energy needs, depending on the purpose of the energy production unit. In fact, the gas turbine can offer a wide range of temperature levels and heat fluxes at the exhaust, because a rapid adaptation to the demand is possible by performing a simple bypass of the exchanger and radiator circuits.

Adapting the respective exergy and anergy levels for a gas turbine allows us to increase its *cogeneration efficiency* E_c, which is expressed as a ratio of the recoverable thermal energy E_{th} and the mechanical energy E_m:

$$E_c = \frac{E_{th}}{E_m} \qquad [6.10]$$

The cogeneration efficiency evolves within a range of 2–6 for gas turbines, and therefore it is significantly higher than that of a more limited piston engine, albeit to the detriment of the primary energy consumption.

For a gas turbine, the main source of discharged heat is concentrated within the exhaust which constitutes a technological advantage in cogeneration applications compared to an internal combustion engine, for which the sources of lost thermal energy involve simultaneously the exhaust and cooling circuits, as well as the coolant, oil and supercharged air (Figure 6.14).

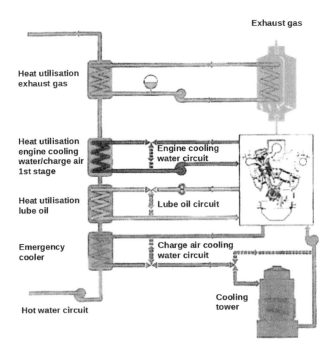

Figure 6.14. *Cogeneration for an internal combustion engine from Boudigues et al. (2001)*

A study was carried out on an installation with intermediate heat exchangers (Figure 6.15), in which a fraction of the thermal energy was taken from the turbine inlet before being returned to the inlet of the combustion chamber (Boudigues et al. 2001). The performance of the cycle changed after simply adjusting the quantity of the air passing through the exchanger, and calculations suggested a possible increase in the thermal efficiency of 20%, bringing it from 0.4 to 0.482. The recovered thermal to generated mechanical energy ratio increases by 40% when the temperature increases linearly from 500 to 900 K.

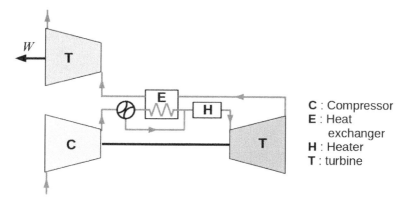

Figure 6.15. *Cycle with an intermediate heat exchanger from Boudigues et al. (2001)*

6.2.4. *Complex depressurized cycle*

The *depressurized cycle* (Figure 6.16), also called the *reverse cycle*, does not involve any prior compression and thermal energy is supplied at the ambient pressure. The hot gases are expanded at a pressure lower than that of the atmosphere, and thermal energy can then be extracted before the cooled gases are recompressed, returning to the ambient pressure p_A. A significant gain in efficiency can be obtained from this cooled compression, which can also supply an external circuit with thermal energy.

This depressurized cycle concerns machines with moderate power levels in particular, and it is observed that the pressure level at the end of the expansion of such a cycle is to be lower than that of a conventional expansion by definition. A priori, this results in a reduction in the density, which penalizes the power and the efficiency of the machine. This drawback is mitigated by increasing the size of the blades, which makes it possible to have the necessary flow rates while reducing the losses generated by the thicknesses of boundary layers on the wall of the airfoils. The rotational speed can thus be reduced by a factor of 3, which makes it possible to find a suitable performance level (Boudigues et al. 2001).

Figure 6.16 illustrates the schematic diagram applied to an automobile engine. The variable geometry turbine and the compressor operate at constant power. The calculation is done for an exhaust temperature of $T_B = 1,200$ K before the turbine. The pressure at the exit of the turbine is equal to 0.069 bar and the temperature is 980 K. The recovery of thermal energy from the outlet of the turbine can be considered, and additional mechanical energy production realized on the rotor of the turbomachinery increases progressively from 2 to 8% of the energy generated on the crankshaft as a function of the load (Descombes and Boudigues 2009).

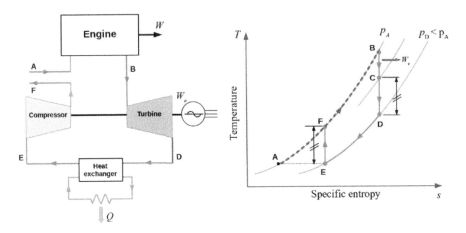

Figure 6.16. *Depressurized cycle from Descombes and Boudigues (2009)*

6.2.5. *Complex over-expansion cycle*

The over-expansion cycle, patented by ONERA (Descombes and Boudigues 2009; Deligant et al. 2012), involves a preliminary compression followed by a heat transfer at constant pressure (Figure 6.17). The combustion gas is expanded to a pressure level below the ambient pressure which is determined by the characteristics of the turbine and compressor. It is then cooled by an exchanger, returning it to the ambient pressure. A significant gain in efficiency can be achieved via a cooled compression which can also supply an external circuit with thermal energy.

Figure 6.18 shows a cogenerating supercharged unit that operates with an over-expanded cycle. The turbocharger is particular in that its operational characteristics are invariant of the mass flow, pressure and speed, regardless of the operating point of the engine. This is achieved from the combined use of a variable geometry turbine along with an appropriate bypass for the supercharged air mass flow placed downstream from the compressor.

Energy Recovery – Waste Heat Recovery 213

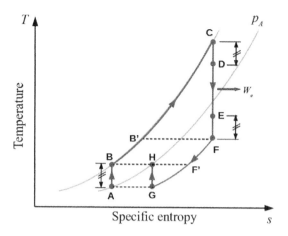

Figure 6.17. *Over-expansion cycle from Deligant et al. (2012)*

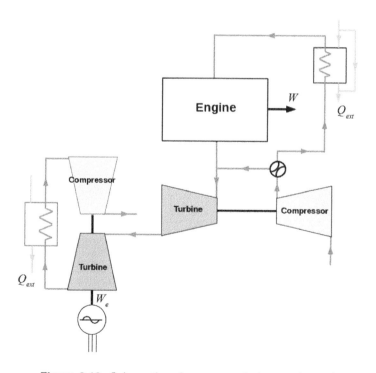

Figure 6.18. *Schematics of a cogenerated supercharged unit with a gas turbine from Deligant et al. (2012)*

Figure 6.19 illustrates the case of a cogeneration unit for the simultaneous generation of electrical and thermal energy. The geometry of the turbocharger is variable, and operates under constant aero-thermodynamic conditions over a wide engine operational range. A bypass is placed downstream from the compressor to re-inject a fraction of the flow preheated by the exhaust gases upstream from the turbine. The calculation is carried out with a supercharging pressure and temperature equal to 2.54 bar and 400 K, respectively. The efficiency of the compressor equals 0.83, and the exhaust temperature is fixed at 1,200 K.

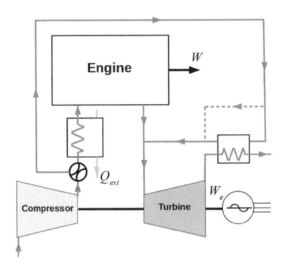

Figure 6.19. *Generation of additional mechanical and thermal energy from Deligant et al. (2012)*

6.2.6. *Conclusion*

We have shown that the aim of the depressurized and over-expansion cycles is to increase the supercharging rate, all while increasing the efficiency. For this, we increase the heights that the compressor and turbine blades, making it possible to lower the rotational angular velocities. A comparative calculation of the depressurized, over-expanded and exchanger cycles is illustrated in Table 6.1, taking into account losses from aerodynamical and mechanical friction.

For a power of 50 kW and a heat exchanger efficiency of 0.85, the values of the optimal compression enthalpy, the mass flow rate, the air–fuel ratio and the angular velocity of the power turbine are stated.

Cycle type	Enthalpy [kJ/kg]	Mass flow rate [kg/s]	Efficiency	Air–fuel ratio	Rotational velocity [rpm]
Exchanger	218	0.229	0.409	0.0127	64,000
Depressurized	225	0.194	0.414	0.0147	31,000
Over-expanded	222	0.193	0.431	0.0143	46,000

Table 6.1. *Comparative performances of cogeneration thermodynamic cycles*

The concept of turbocharging is a recognized practice which lets us substantially increase the torque and power of an automobile or industrial transportation engine. Strictly speaking, this is not cogeneration, but the fact remains that the supercharged turbine takes thermal energy discharged from the exhaust and uses it to produce mechanical energy which will drive the compressor.

The calculations show that these cycles are, a priori, feasible on an industrial scale and can generate a potential improvement of the order of 5–10% in energy efficiency, but they clearly give rise to technological complexities, for which the additional cost remains yet to be quantified. A possible avenue towards energy recovery in automobile engines that allow for additional electricity production is realistic.

We see that the significant variability of the flow rates in automobile engines creates strong constraints, because such an engine operates with a partial load whose mass flow rate typically is in the range of a few tens of g/s. The most obvious case is the use of a vehicle subjected to a depollution cycle where the levels of PME are then very reduced.

6.3. Micro-cogeneration

6.3.1. *Introduction*

The significant energy and environmental disorder leads us to consider an energy policy directed towards three simultaneously major concerns: working against climatic disorder, security of primary energy supplies and the preservation of health and the environment. Micro-CHP is a *technique used for the simultaneous production of electrical and hot and cold thermal energy in moderately sized power units*. It is well suited to individual and collective housing, as well as to the tertiary and hospital sectors. It concerns all types of heat engines with both internal and external combustion, gas micro-turbines, in addition to fuel cells. This combined production makes it possible therefore to improve the utilization rate of the primary energy for a given energy unit (Onovwiona and Ugursal 2006).

We can only strongly recommend that the reader refers to the numerous annual micro-cogeneration days, which have been organized at the Conservatoire National des Arts et Metiers for the past 20 years or so and regularly disseminate the scientific advances in both micro- and mini-cogeneration. Also each year, they update the complex normative developments dedicated to this evolving subject, as well as condition the thermo-economic regulations for the recovery of waste energies. These days are organized by the Association Technique Énergie Environnement (atee.fr), in addition to French industrial and academic specialists. For example, see the portal: https://events.femto-st.fr/Journees-Cogeneration/.

6.3.2. *Classification*

Though not unique, the standard classification typically states that micro-CHP concerns units whose power rarely exceeds 50–80 kW, while mini-CHP concerns those whose power lies below 215 kW. These cogeneration installations correspond to applications for individual housing, and collective and tertiary sectors with regard to their size. We use the term "small" cogeneration for such installations between 215 kW and 1 MW, and 1 and 12 MW for "medium" cogeneration. "Big" cogeneration applies to those with powers greater than 12 MW. To compare the performance of different systems, the following definitions are used:

– τ_{elect} is the electrical efficiency, and measures the ratio between the electrical energy output and the fuel energy input;

– τ_{therm} is the thermal efficiency, and measures the ratio between the useful thermal energy output and the fuel energy input;

– τ_{coge} is the overall cogeneration efficiency, which expresses the ratio between the sum of the electrical energy with the useful thermal energy output over that with the fuel energy input.

Of course, these are not thermodynamic efficiencies in the sense of the second principle, since the same unit of energy is attributed both to the useful thermal and electrical energies in these definitions, with no reference to Carnot's postulate. These criteria simply measure the usage rate of the primary energy from the fuel input, while the significant ratio of electricity to heat is the subject of regulations in terms of needs distributed between heat and electricity in cogeneration.

6.3.3. *Internal combustion engines*

The power unit commonly used in power generation plants with a modest power output is that of an internal combustion engine. The engine works at a constant rate and offers a high efficiency which is close to 50% (Abusoglu and Kanoglu 2009), and a substantial part of the fuel energy rejected in the form of heat is partially regenerated. Residual energy in the exhaust and cooling systems is often used to produce hot water and steam.

A CHP positive displacement engine is a complex system, requiring a generator and additional boilers to heat the water to be used in the local heating system (Mebarkia et al. 2017, 2019). The disadvantage of engines driven by a cogeneration system is that the engine must be placed close to the heated object (building, factory, city, etc.). The operating cycle of a cogenerated engine is shown in Figure 6.20. Benelmir and Feidt (1998) and Kanoglu and Dincer (2009) report that the overall efficiency of a combustion engine, which is powered by such a cogeneration system, ranges from 75% to 80%.

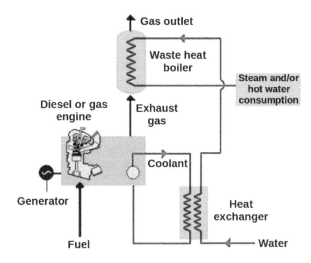

Figure 6.20. *Schematic diagram of a cogenerated internal combustion engine (ATEE source)*

Reciprocating internal combustion engines are well suited for small- or large-scale cogeneration because they are robust and their technology is recognized and been proven to be reliable (Dolz et al. 2012). They require regular maintenance and are available in a wide electrical power range, from 10 kW to 10 MW for different fuels. These engines are classified into two large families: spark-ignition engines and compression ignition (or gas) engines. Diesel engines are mainly used for high-power stationary cogeneration and can run on diesel or heavy fuel oil. Spark-ignition engines are well suited to moderate-power cogeneration uses and can produce pressurized water at 160°C, and can run on gasoline, natural gas, propane gas or biogas.

The primary energy utilization rate ranges between 75 and 85% depending on the fuel load level. In general, the heat engine drives a synchronous generator at a constant rate, so that the frequency of the current generated coincides with that of the electrical grid. In general, the cogeneration installation is appropriately sized to optimize the electrical efficiency. This electrical efficiency remains more or less constant for loads between 100% and 75%, before then decreasing. The heat produced cannot be fully recovered, and recovery primarily involves the exhaust gases, the engine cooling water and, to a lesser extent, the oil cooling circuits.

Periodic maintenance should be scheduled every 500 or 2,000 hours to change the oil, coolant and spark plugs, and top-end overhaul (cylinder head and turbocharger to be changed) should be scheduled every 12,000 or 15,000 hours. Finally, a major overhaul is necessary every 24,000 or 30,000 hours to replace the sealant rings and the crankshaft bearings. The main pollutants emitted by these engines are nitrogen oxides NO_x, carbon monoxide CO and volatile organic compounds, as well as other particulate matter in the case of diesel engines since they run on fuel oil.

A heat engine with a high efficiency naturally leads to exponential emissions of NO_x due to their high combustion temperatures. However, it is possible to adjust the engine to favor lower NO_x emissions albeit to the detriment of efficiency. This results in a greater production of CO and unburnt gases, so that it is then necessary to use an oxidation catalyst (e.g. a catalytic converter) to convert them into carbon dioxide CO_2 and water.

In motor vehicles, waste heat recovery from exhaust gases is an effective technology though at the compromise of increased complexity and significant bulkiness (Glavatskaya et al. 2012; Danel et al. 2015). Higher exergy is contained in the exhaust gases than in the cooling system. The Rankine cycle and its derivatives offer a high potential, as the thermal efficiency of the system can be increased by an additional 8–10%.

6.3.4. *Gas micro-turbines*

Gas micro-turbines operate on the same general principle as larger power turbines. However, due to their size, they are subject to specificities when compared to large installations (Wang et al. 2005). For low powers, the mass flow rate in the machine is low. Furthermore, axial machines are no longer suitable so radial machines are then used, similar to those found in automobile turbochargers (Pilavachi 2002; Romier 2004).

For this type of application where cost is an important parameter, the technology must remain unsophisticated. Cooled blades are therefore not possible. The maximum allowable temperature determined by the resistance of the materials is approximately 800–900°C. As a result, the optimal compression rate as well as the performance drops sharply (Wang et al. 2005; Traverso et al. 2006). In a simple cycle, electricity would be produced at an efficiency of less than 15%. To improve energy conversion, a cycle with an exhaust gas heat recovery system is opted for (Figure 6.21) (Kanoglu and Dincer 2009).

This heat recovery exchanger remains a critical point in more ways than one. It is sensitive to fouling and must withstand high temperatures. Increasing its efficiency, which directly affects the overall electrical efficiency, consequently leads to an increase in volume and hence its cost, which constitutes approximately 25% of the price of a gas micro-turbine. As the volumetric compression ratio is low (approximately 4), a single compression stage (or sometimes two) is usually sufficient, and expansion is still able to take place in a single stage since expanding a fluid is easier to do than compressing it (Traverso et al. 2006).

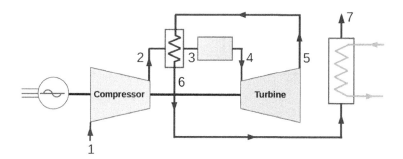

Figure 6.21. *Schematic diagram of micro-cogeneration system with a gas turbine and heat recovery exchanger 2–3*

Thus, the *heat recovery cycle* is composed of the following phases (Figure 6.21):

– *adiabatic compression* 1–2, of the intake air;

– *compressed air heating* 2–3, in a heat exchanger, which recovers some of the exhaust gas energy;

– *constant pressure combustion* 3–4, in the combustion chamber;

– *adiabatic expansion* 4–5, in the turbine.

There is still some energy in the gases (at 250 or 300°C) that leaves the machine. We can therefore insert another exchanger 6–7 which can *produce heat* for a space heating application, for example. Due to the low mass flows and powers, the geometry of the machines is miniaturized. However, the smaller the machine then the higher the rotational velocity for the same limited peripheral rotor velocity, so for the same specific energy (J/kg of fluid) that is exchanged on the shaft. While large gas turbines have a rotational velocity of 3,000 rpm and are directly coupled to the network at 50 Hz, small turbomachines used in mini-cogeneration have speeds of the order of 75,000–120,000 rpm for powers below 100 kW. As the rotational velocity is very high, it is also necessary to provide a gearbox or a high-speed alternator, along with suitable power electronics (Kim and Hwang 2006).

To overcome this problem, some gas micro-turbines may have two shafts: a power turbine (also called a free turbine) and a turbine linked to the compressor (Figure 6.22), with the coupling between the two parts of the machine only done by the fluid (point 4). Among other things, this allows the power turbine to rotate at a lower velocity, which is more compatible with other electrical machines. On the other hand, it also helps to promote certain operational flexibility with a partial load (Kim and Hwang 2006).

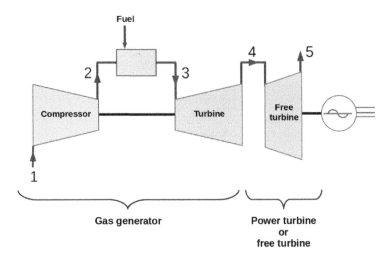

Figure 6.22. *Twin-shaft gas turbine*

To summarize, for all the reasons mentioned above, gas micro-turbines used in cogeneration systems generally have the following characteristics:

– a centrifugal compressor (one or two stages);
– a low compression ratio (approximately 4);
– a very high velocity (25,000 rpm for 500 kW, 75,000 rpm for 100 kW);
– a high-speed alternator and power electronics to convert high-frequency currents into direct currents, and then to mains-frequency currents;

– a heat recuperator for the exhaust gases that preheats the air before combustion (after the compressor);

– recovery of remaining exhaust heat energy for cogeneration.

Gas micro-turbines can operate with a wide range of liquid or gaseous fuels, but solid fuels are excluded (Pilavachi 2002). As a result, the use of biomass can only be realized through biogas, bio-fuel and wood gas, making things more complicated overall. To allow for the use of solid fuel, a heat exchanger – instead of the combustion chamber – would theoretically be possible, as was investigated in an 80 kW micro-turbine (Traverso et al. 2006). However, the major obstacle lies in the need for a high-temperature exchanger, or otherwise only low electrical efficiencies will be obtained. The fouling of this exchanger is also a point to keep in mind (Wang et al. 2004).

Gas micro-turbines emit fewer pollutants NO_x and CO than internal combustion engines because they use continuous combustion, even if at partial load the emissions of unburnt matter remain high (Pilavachi 2002; Romier 2004). Maintenance requirements are low (every 8,000–11,000 hours), compared to those for reciprocating internal combustion engines, but they require the intervention of specialist personnel that generates additional operating costs.

6.3.5. *Fuel cells*

Fuel cells convert the enthalpy of a fuel directly into an electric current without the need for rapid combustion from an oxidation–reduction reaction (Stevens et al. 2000). There are two main families of so-called acid or alkaline batteries. While in a conventional electrical conductor, electricity is conveyed by the movement of electrons (external circuit), in the core of the batteries, this is not possible for them and instead it is the ions crossing the electrolyte. In an acid battery, it is the proton H^+ which is involved and the half-reactions are as follows:

Anode: $\qquad 2\,H_2 + 4\,H_2O \longrightarrow 4\,H_3O^+ + 4\,e^-$

Cathode: $\qquad O_2 + 4\,H_3O^+ + 4\,e^- \longrightarrow 6\,H_2O$ [6.11]

On the other hand, in an alkaline battery, the hydroxide anion OH^- passes through the electrolyte with the following half-reactions:

Anode: $\qquad 2\,H_2 + 4\,OH^- \longrightarrow 4\,H_2O + 4\,e^-$

Cathode: $\qquad O_2 + 2\,H_2O + 4\,e^- \longrightarrow 4\,OH^-$ [6.12]

Note that, when hydrogen is used as a fuel, the overall reaction always results in the oxidation of hydrogen:

$$2\,H_2 + O_2 \longrightarrow 2\,H_2O \qquad [6.13]$$

A fuel cell makes it possible to separate the reactions between the anode and the cathode and collect the electrons into an external circuit. The dissociation of the fuel and the recombination of the ions require a catalytic action on the part of the electrodes. Ultimately, the water formed must either be discharged or reused in the battery system. The thermodynamic irreversibilities and the ohmic resistance of the battery cell generate heat which must also be discharged, and which can thus be used for heating.

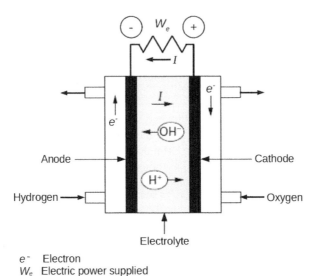

e^- Electron
W_e Electric power supplied

Figure 6.23. *Schematic diagram of a hydrogen–oxygen fuel cell*

The hydrogen used as one of the possible fuels can be produced from many different sources such as natural gas, propane or coal, from steam reformation, or even (and preferably) from water electrolysis. Reforming can take place outside of the cell (in low-temperature cells, such as PAFC, AFC or PEMFC) or, even better, inside of the cell (in high-temperature cells, such as MCFC or SOFC), increasing the efficiency and lowering manufacturing costs.

These reactions occur at 700°C in the presence of a catalyst (Ni), leading to a gaseous mixture rich in hydrogen. The presence of an external reformer increases the complexity of the system and reduces energy efficiencies (reformer efficiency from 70 to 80%). There are different types of fuel cells currently at different stages of development, with the main difference being in the type of electrolyte used and the conducting ions:

– PAFC: phosphoric acid fuel cells. They are most often used developed for terrestrial applications and the first to be commercialized. They operate at approximately 200°C, which is beneficial for cogeneration.

– AFC: alkaline fuel cells. The liquid electrolyte is aqueous KOH. They are characterized by a low operating temperature from 60 to 80°C. Level of advancement: prototype (10 kW) and unit production.

– PEMFC: proton-exchange membrane fuel cell. The electrolyte is a polymer membrane proton conductor. They also function at low temperatures typically between 80 and 90°C with very pure hydrogen. Level of advancement: prototype (250 kW).

– SOFC: solid oxide fuel cell (ZrO_2 and Y_2O_3). The transport ion is O_2^-; they operate at high temperatures of approximately 950 or 1,000°C. State of advancement: prototype (100 kW).

– MCFC: molten carbonate fuel cell. Li_2CO_3 and KCO_3 in a $LiAlO_2$ matrix. The transport ion is CO_2^{--}; they operate between 600 and 700°C. This technology is still in the development phase. Prototype 2 MW.

– DMFC: direct methanol fuel cell. State of progress: research and development. Industrialization was planned in around 2015.

The electrical efficiency varies according to the type of battery and is usually between 30 and 45%. Fuel cell efficiency improves with partial loads and heat is recovered from both the reformer and the cell. The high-temperature levels attained in the MCFC and SOFC fuel cells make it possible to envisage additional electricity production, using a steam turbine, for example. Feedback remains to be consolidated but, apart from auxiliary systems such as the fuel pump and the fan, fuel cells have no moving parts and therefore should require little maintenance.

The main source of harmful emissions comes from the reforming process which requires combustion. If the temperature is kept below 1,000°C, the emissions of NOx remain limited and this temperature remains sufficient to oxidize the CO.

6.3.6. *Thermoelectricity*

Thermoelectricity is a phenomenon that arises from the *Seebeck effect*, discovered by Thomas Johann Seebeck in 1821. Electrical energy is produced directly from a heat flow through an assembly of two dissimilar semiconductor metals (Le Blanc et al. 2014), and these two materials must have different electronic conductivities (Figure 6.24).

The main advantage of this technology is the absence of mechanical parts. Due to its simplicity, the automotive application of thermoelectricity is something being actively developed, though its efficiency is lower than that of other thermodynamic cycles in addition to its cost being higher. Car manufacturers have stepped up their efforts on the subject, but new European directives have prohibited any thermal energy production machines for the decade 2020–2030. A possible optimization initiative would be to recover waste heat from exhaust gases via a thermoelectric conversion that exploits the Seebeck effect to improve vehicle fuel efficiency and further reduce CO_2 emissions.

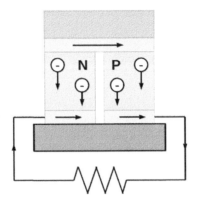

Figure 6.24. *Thermoelectric generator*

6.3.7. *Thermoacoustics*

An innovative conversion process is that which uses thermoacoustic technology, in which acoustic energy is converted into thermal energy. In general, thermoacoustics is the coupling phenomenon between sound wave propagations described by the motion, pressure and temperature oscillations, and the oscillatory heat transfer between the compressible fluid (gas in most cases) traveling within a small channel and the neighboring solid boundaries. The development of this technology, which is the most recent among heat engines, began in the 1980s. It consists of amplifying an acoustic wave with a heat flux induced by a temperature gradient (Haddad et al. 2014). The acoustic wave energy is converted into electrical energy using a microphone or a piston connected to a linear electrical alternator (Figure 6.25). It can be used either as a motor or as a refrigerator. As with thermoelectricity, a big advantage of this technology is the very moderate number of mechanical parts, guaranteeing robustness and reliability of the device.

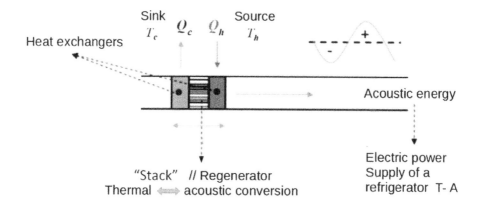

Figure 6.25. *Schematic diagram of a thermoacoustic engine*

6.3.8. *"Rankinized" cycles*

The Rankine and Hirn cycles are used in steam engines, which are the oldest type of heat engine. In the 1st century AD, the Hero of Alexandria discovered the possibility of converting heat into work (Thurston 1983). From the 18th century onward, steam engines were used industrially and, in 1781, James Watt's steam engine established the current form of this cycle.

Historically, *water* was the working fluid though *other fluids* are also used, many being organic in which case the cycle is called the *organic Rankine cycle* (ORC). In a Rankine cycle, the fluid is pumped into a heat exchanger where it is vaporized into steam, which is then expanded in a turbine or some other expander (historically a piston machine), before then being condensed in a condenser. In the Hirn cycle, the difference is that the steam is superheated once it has evaporated, but otherwise the rest of the cycle is identical (Figure 6.26).

It is the most widely used technology in the industry, as well as the best studied for any prospective automotive waste heat recovery applications. The Rankine–Hirn cycle has a high potential for energy recovery due to a low back-pressure effect, and the highest efficiency that it can reach is over 10% (Wang et al. 2011; Sprouse and Depcik 2013). In addition, Bianchi and De Pascale (2011) have shown that the Rankine cycle offers a better performance than other technologies. The variation in the thermodynamic parameters of the working fluid is shown as a $T - s$ diagram in Figure 6.27.

226   Thermodynamics of Heat Engines

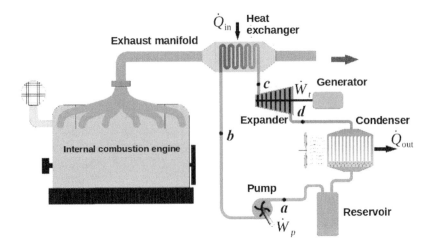

Figure 6.26. *Exhaust heat recovery in an engine using the Rankine cycle*

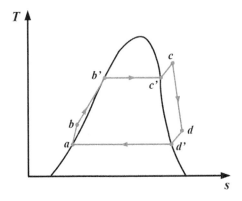

Figure 6.27. *Representation of the Rankine cycle on a T–s entropy diagram*

The waste energy in the exhaust and cooling systems is used to produce hot water and steam. A cogenerating engine is a complex system that requires a generator and boilers to heat the water to be used in the local heating system. The disadvantage of engines driven via a cogenerating system is that the engine must be placed within the vicinity of the object to be heated (building, factory, city, etc.). The operating cycle of a cogenerated engine is shown in Figure 6.28. The overall efficiency of such a cogeneration system can be calculated using the relation:

$$\eta_{cog} = \frac{\dot{W}_{net} + \dot{Q}_{heat}}{\dot{Q}_{fuel}} \quad [6.14]$$

Benelmir and Feidt (1998) and Kanoglu and Dincer (2009) claim that the overall efficiency of a combustion engine that is powered via a cogeneration system is between 75% and 80%.

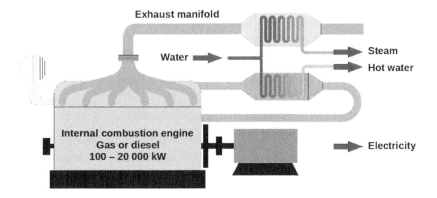

Figure 6.28. *Cogenerated internal combustion engine*

Polygeneration is an interesting prospect as a way to increase the thermal efficiency of engines. Engine cogeneration technology can increase thermal efficiency by up to 75–80%. This technology is appropriate for an engine that powers a power plant, since the energy lost in the exhaust and cooling system produces thermal energy which can then be used in a heating system.

In motor vehicles, waste heat recovery from exhaust gases is the most efficient technology (Glavatskaya et al. 2012; Danel et al. 2015). Not surprisingly, the exhaust gases contain a much greater exergy than in the cooling system where temperatures are lower. The energy recovered from the vehicle is converted into mechanical or electrical energy, increasing the efficiency of the system by a few percent.

6.4. Conclusion

Cogeneration systems are very interesting from the point of view of efficiency. However, it is important to correctly gauge their dimensions in accordance to their heat and electricity requirements. The responsible uses of energy and environmental protection are two reasons why it is strongly recommended to get their dimensions correct for their heat requirements. This avoids any massive heat rejection into the environment, and then any electricity generated is considered to be a byproduct that is incorporated into the electrical grid.

Cogenerated internal combustion engines are the most reliable technology, but require check-ups more frequently and periodically than other systems, such as gas micro-turbines. Furthermore, they emit more CO, NO_x and particulates than other competing technologies. The Stirling engine should not require too much maintenance and also fuel cells have only a few moving parts; hence, they also have the potential for low maintenance requirements.

Cogeneration plants that use renewable energy face additional constraints and pose specific problems, which can be summarized as follows:

– the use of biogas requires a methanizer or an anaerobic digester;

– non-esterified vegetable oils which can be used as a biofuel are not authorized in France (except in special cases);

– the gasification industry remains to be properly established, requires substantial installations and is not very widespread.

The vast majority of boiler manufacturers are currently interested in using cogeneration in applications for the tertiary sector, as well as for collective and individual housing. In France, after a demonstration campaign by Gdf Suez with approximately 40 boilers installed in private homes, these systems are now being commercialized. The main projects are those concerning the industrial and tertiary sectors, since these technologies are still expensive for individuals. Regulatory framework could evolve further to include additional incentives, as a bid to reduce the connection costs to the electrical grid.

6.5. Perspectives

High-power electrical power plants CCGT ("Combined Cycle Gas Turbine") that use combined gas and steam turbine cycles and have been in operation in France for more than 15 years contribute to the goal of reducing the consumption of fossil fuel resources thanks to the improved efficiency that they bring. Among these power plants are the following:

– The DK6 power plant in Dunkirk (Nord), commissioned in 2005 with two 395 MW units (gas turbine: 165 MW; steam turbine: 230 MW) and runs on natural and blast furnace gas.

– The EDF combined cycle power plant in Martigues (Bouches-du-Rhône), commissioned between 2012 and 2013 and built reusing parts of the old facility with two 465 MW units that run on natural gas.

– The new generation combined cycle power plant in Bouchain (Nord), with a capacity of 605 MW running on natural gas, and with a high degree of operating flexibility (production ramp-up to 50 MW/minute) and an efficiency of almost 62%. The high-efficiency value makes it possible to estimate an overall reduction of approximately 55% in CO_2 emissions, compared with conventional coal-fired power plants.

Cogeneration is an opportunity for significant savings in primary energy resources. By simultaneously producing two forms of energy within the same power plant, typically electrical and heat energy, coming the same primary energy fuel source, either fossil (natural gas, etc.) or renewable (biomass, etc.). Cogeneration, which consists, for example, of using heat released by the drive of an alternator by a heat engine or by a turbine when this resource is otherwise unused, makes it possible to achieve efficiencies exceeding 85%, according to the first principle of thermodynamics.

Cogeneration systems are generally classified according to their electrical power produced, ranging from less than 36 kVA in the case of micro-cogeneration, to more than 12 MW in the case of large cogeneration.

Industrial cogeneration plants (food industry, paper mills, chemicals, etc.), with power ratings typically between 25 and 90 MW, are mainly used for producing simultaneously electrical energy and steam. District heating applications are usually within the same power range.

Domestic applications that harness micro-cogeneration are still not widely used in France. Generally, they use Stirling engines or internal combustion engines, while those with fuel cells are still at the testing stage. The use of Ericsson engines is the subject of other research and development work.

Finally, a further step can be taken with tri-generation systems which produce a third type of energy, typically in the form of refrigeration.

6.6. References

Abusoglu, A. and Kanoglu, M. (2009). Exergetic and thermoeconomic analysis of diesel engine powered cogeneration: Part 2 – Application. *Applied Thermal Engineering*, 29, 242–249.

Benelmir, R. and Feidt, M. (1998). Energy cogeneration systems and energy management strategy. *Energy Conversion and Management*, 39, 1791–1802.

Bianchi, M. and De Pascale, A. (2011). Botoming cycles for electric energy generation: Parametric investigation of availabler and innovative solutions for the exploitation of low and medium temperature heat sources. *Applied Energy*, 88, 1500–1509.

Boudigues, S., Descombes, G., Neveu, P., Prévond, L. (2001). Optimisation d'unités cogénérées de production d'énergie par turbines à gaz et moteurs. *Heat Powered Cycles Conference (HPC'01)*, Paris, 5–7 September 2001.

Danel, Q., Périlhon, C., Lacour, S., Punov, P., Danlos, A. (2015). Waste heat recovery to a tractor engine. *Energy Procedia*, 74, 331–343.

Deligant, M., Descombes, G., Chiriac, R. (2012). Analyse de cycles thermodynamiques complexes de poly–génération. *Termotehnica*, 1, 24–29.

Descombes, G. (2003). Transferts de masse et de chaleur dans les moteurs thermiques et récupération d'énergie. HDR, Pierre and Marie Curie University (UPMC), Paris.

Descombes, G. and Boudigues, S. (2009). Modelling of waste heat recovery for combined heat and power applications. *Applied Thermal Engineering*, 29, 2610–2616.

Descombes, G., Jullien, J., Magnet, J.-L., Murat M. (1999). Évolution technologique des moteurs diesel industriels. *Revue française de mécanique*, 4.

Dolz, V., Novella, R., García, A., Sánchez, J. (2012). HD Diesel engine equipped with a botoming Rankine cycle as a waste heat recovering system. Part 1: Study and analysis of the waste heat energy. *Applied Thermal Engineering*, 36, 269–278.

Glavatskaya, Y., Podevin, P., Lemort, V., Shonda, O., Descombes, G. (2012). Reciprocating expander for exhaust heat recovery rankine cycle for a passenger car application. *Energies*, 5, 1751–1765.

Haddad, C., Périlhon, C., Danlos, A., François, M.-X., Descombes, G. (2014). Some efficient solutions to recover low and medium waste heat: Competitiveness and the thermoacoustic technology. *International Conference on Technologies and Materials for Renewable Energy, Environment and Sustainability, TMREES 14*, Beirut, 10–13 April 2014. *Energy Procedia*, 50, 1056–1069.

Kanoglu, M. and Dincer, I. (2009). Performance assessment of cogeneration plants. *Energy Conversion and Management*, 50, 76–81.

Kim, T.S. and Hwang, S.H. (2006). Part load performance analysis of recuperated gas turbines considering engine configuration and operation strategy. *Energy*, 31, 260–277.

Le Blanc S., Yee S.K., Scullin M.S., Dames C., Goodson K.E. (2014). Material and manufacturing cost considerations for thermoelectrics. *Renewable and Sustainable Energy Reviews*, 32, 313–327.

Mebarkia, M., Louafi, M., Zoubir, A. (2017). Statistical design of experiments as a tool for investigation for the influence of porous coating on the critical heat flux. *International Conference on Technologies and Materials for Renewable Energy, Environment and Sustainability, TMREES 17*, Beirut, 21–24 April 2017. *Energy Procedia*, 119, 1003–1011.

Mebarkia, M., Louafi, M., Zoubir, A. (2019). Energetic transition within thermal machines and co-generation: effect of mass flux on critical heat flux. *Progress in Industrial Ecology – An International Journal*, 13(2), 111–123.

Milkov, N., Punov, P., Evtimov, T., Descombes, G., Podevin, P. (2014). Energy and exergy analysis of an automotive direct injection diesel engine. *BulTrans–2014, Sozopol*, 149–154.

Onovwiona, H.I. and Ugursal V.I. (2006). Residential cogeneration systems: Review of the current technology. *Renewable & Sustainable Energy Review*, 10, 389–431.

Pilavachi, P.A. (2002). Mini- and micro-gas turbines for combined heat and power. *Applied Thermal Engineering*, 22, 2003–2014.

Pluviose, M. (2009). *Conversion d'énergie par turbomachines*. Collection Technosup, Ellipses, Paris.

Punov, P., Lacour, S., Perilhon, C., Podevin, P. (1993). Possibilities of waste heat recovery on tractor engines. *Proceedings of Scientific Conference BulTrans-2013*, Sofia, Bulgaria, 7–15.

Romier, A. (2004). Small gas turbine technology. *Applied Thermal Engineering*, 24, 1709–1723.

Sprouse, C. and Depcik, C. (2013). Review of organic Rankine cycles for internal combustion engine exhaust waste heat recovery. *Applied Thermal Engineering*, 51, 711–722.

Stevens, P., Novel-Cattin, F., Hammou, A., Lamy, C., Cassir, M. (2000). Piles à combustible. *Techniques de l'Ingénieur*, D 3 340, 1–28.

Thurston R.H. (1883). *A History of the Growth of the Steam-engine*, 3rd edition, Kegan Paul, Trench & Co's Publications.

Trapy, J.D. (1981). Heat transfer in internal combustion engine. *Revue générale de thermique*, 233, 385–390.

Traverso, A., Massardo, A.F., Scarpellini, R. (2006). Externally fired micro-gas turbine: Modelling and experimental performance. *Applied Thermal Engineering*, 26, 1935–1941.

Wang, W., Cai, R., Zhang, N. (2004). General characteristics of single shaft microturbine at variable speed operation and its optimization. *Applied Thermal Engineering*, 24, 1851–1863.

Wang, W., Chen, L., Sun, F., Wu, C. (2005). Power optimization of an endoreversible closed intercooled regenerated Brayton cycle. *International Journal of Thermal Sciences*, 44, 89–94.

Wang, T., Zhang, Y., Peng, Z., Shu, G. (2011). A review of researches on thermal exhaust heat recovery with Rankine cycle. *Renewable and Sustainable Energy Reviews*, 15, 2862–2871.

List of Authors

François COTTIER
MTU Aero Engines
Munich
Germany

Georges DESCOMBES
CNAM
Paris
France

Bernard DESMET
INSA – HdF
Université Polytechnique
Hauts-de-France
Valenciennes
France

Michel FEIDT
LEMTA
Université de Lorraine
Vandoeuvre-les-Nancy
France

Mohamed MEBARKIA
Larbi Tebessi University
Algeria

Yannick MULLER
MTU Aero Engines
Munich
Germany

Index

A, B

aerodynamics, 70, 107, 110, 132
air, 14, 15, 19, 20, 25, 26, 28, 30, 33, 43–47, 51, 61–63, 66, 69, 70, 72, 74, 75, 78, 80–83, 86–91, 96–98, 100, 101, 103, 105, 111–118, 120, 122, 123, 125–132, 134, 135, 137, 138, 140, 142–145, 150, 152–155, 164, 165, 170–172, 176, 177, 179, 187–193, 197–199, 205, 210–212, 219, 221
air–fuel ratio, 214, 215
airfoil, 90, 110, 113, 130, 211
authorization, 107
balance, 82, 163, 197–199, 202–204

C, D

chemical kinetics, 155, 156, 164, 177
classification, 72, 206, 216
coefficient, 15, 18, 22, 55–57, 67, 80–82, 141, 142, 147, 153, 157, 158, 162, 171, 177, 188, 190–192
cogeneration, 14, 15, 67, 205, 207–210, 214–218, 220, 221, 223, 227–231
combustion, 6–8, 21, 23–28, 30–34, 37, 39–45, 47, 49–54, 56–63, 66–68, 70, 72–74, 76, 78, 79, 83, 86–89, 91, 93, 95, 97, 98, 100, 103–105, 108, 109, 111, 116, 119–121, 127, 129, 133–135, 137–156, 160, 161, 163–166, 169–174, 176, 177, 179, 188, 192, 194, 195, 197, 198, 203, 206, 210–212, 216–219, 221, 223, 227–229, 231
chamber, 74, 76, 79, 86–89, 91, 93, 95, 97, 98, 100, 105, 108, 109, 111, 119–121, 127, 129, 137, 140, 188, 211, 219, 221
constant
pressure, 146, 149, 219
volume, 49, 66, 144, 145, 147, 150
external, 6–8, 179, 195, 203, 216
internal, 21, 23–25, 43, 44, 51, 61–63, 66–68, 70, 72, 104, 135, 139, 177, 192, 194, 197, 198, 203, 206, 210, 217, 221, 227–229, 231
products, 44, 45, 59, 91, 145, 149, 155, 160, 161, 165, 166
composition, 13, 42, 45, 87, 133, 142, 153, 156, 160, 161, 163
compression, 6, 23, 24, 27, 28, 30–33, 36–41, 46, 48–50, 56, 57, 61, 63, 66, 75, 87, 89, 93, 95, 96, 98, 100, 101, 103, 104, 109, 115–118, 139–141, 181, 182, 188–190, 194, 195, 211, 212, 214, 217, 219, 220
ratio, 24, 27, 28, 66, 75, 100, 101, 109, 116, 139, 140, 219, 220
compressor, 2, 61–64, 67, 74–76, 80, 86–91, 93–101, 103–105, 109–121,

125, 127–130, 187–189, 192, 193, 212, 214, 215, 220, 221
connecting rod, 23, 29, 30
constant, 14–16, 18, 19, 27, 30–32, 34, 37, 39–46, 48–50, 54, 56–59, 63–66, 73, 78, 80, 87, 88, 91–95, 97, 99, 110, 118, 121, 129, 130, 135, 144–150, 152–155, 158–160, 162–164, 167, 168, 170–173, 180–182, 189–191, 198, 212, 214, 217–219
consumption, 61, 72, 78, 81, 85, 130, 132, 205, 210, 228
conversion, 1, 5, 6, 67, 133, 157, 195, 196, 200, 202, 203, 206, 209, 219, 224, 229–231
crankshaft, 23, 54, 57, 58, 200–202, 212, 218
cycle
    Carnot, 9, 10, 104
    combined, 67, 207, 208, 228
    development, 104, 125, 126
    diesel, 39–43
    Ericsson, 104
    Joule–Brayton, 93, 94, 96, 98–103, 109, 115, 121, 126
    limited pressure (mixed), 41, 42
    Miller–Atkinson, 36–38, 43, 66
    Otto (Beau de Rochas), 31
    Stirling, 181, 183, 184
cylinder, 10, 11, 23–32, 34, 35, 37, 39, 50–54, 56–59, 61–65, 67, 135, 137, 139, 140, 179, 180, 184, 186, 188, 189, 191, 197, 218
deflagration, 134
degradation, 5, 6, 36, 114, 200, 203
design, 88, 107, 109–111, 115, 122, 125, 231
diffusion, 5, 118, 136, 137, 140, 165
dissociation, 155, 156, 160, 163, 222
dithermal, 9, 200
drag, 81, 82

E

efficiency
    Carnot, 7, 10, 104, 183, 184, 201
    cogeneration, 209, 210
    mechanical, 63, 64
    thermal, 84, 94, 95, 101, 102, 104, 183, 184, 211, 216, 218, 227
    thermodynamic, 7, 10, 33, 34, 36, 38, 40–43, 50, 91, 128, 129, 193, 216
end of life, 108
endoreversible, 9, 10, 232
energy
    activation, 134, 135, 137, 158
    internal, 1–3, 11, 15, 21, 32, 48, 52, 53, 74, 145, 149, 150, 182, 200
    kinetic, 3, 5, 12, 34, 53, 65, 74, 84, 85, 96, 99, 115, 117, 119, 122, 124, 156
    mechanical, 5, 6, 43, 65–67, 88, 98, 99, 130, 135, 172, 188, 195, 203, 205, 209–212, 215
    potential, 12, 64, 115, 118, 119, 124, 172
    thermal, 5, 6, 67, 115, 134, 165, 180, 183, 195, 198, 200, 205, 206, 210–212, 214–216, 224, 227
engine, 5–9, 23–31, 33, 44, 47, 51, 52, 54, 57–59, 61–63, 67, 68, 70–76, 79, 82, 83, 85, 86, 88–90, 93, 97, 99, 101, 103–105, 107–114, 119, 124, 125, 127–131, 137, 139–141, 143, 179, 180, 184–188, 191–195, 197–205, 207–210, 212, 214, 215, 217, 218, 225–231
    Ericsson, 179, 187, 188, 191, 193–195
    expansion, 188, 191–193
    external combustion, 6–8
    internal combustion, 24, 44, 51, 63, 67, 68, 192, 194, 197, 198, 210, 217, 227, 231
    Stirling, 179, 180, 184–187, 194, 195, 228
enthalpy, 3, 15, 55, 60, 66, 92, 99, 147–149, 152, 153, 160, 167, 170, 174, 200, 204, 205, 214, 215, 221
    formation, 147–149, 151, 173, 175
    reaction, 147, 148
entropy, 7, 9–11, 15, 16, 21, 31, 32, 63, 65, 88, 91–94, 167, 168, 172–174, 181, 200, 202, 204, 226

Index 237

equation of state, 14, 16, 142, 166, 167, 169, 182
equilibrium, 5, 10, 14, 59, 87, 88, 155, 156, 159–163, 168, 181, 186, 204
exergy, 10–14, 21, 65–67, 166–172, 176, 195, 200, 202–205, 210, 218, 227, 231
exhaust, 23, 26, 28–32, 36, 47, 50–53, 59, 61–67, 86, 88, 89, 98, 105, 109, 114, 124, 190–193, 197, 198, 203–207, 210, 212, 214, 215, 217–219, 221, 224, 226, 227, 230–232
extensive quantity, 152, 157
extrados, 80, 123

F

fan, 77, 90, 109, 127, 223
flame, 27, 114, 119, 120, 134–140, 150, 152–155, 170, 171, 177
  diffusion, 136, 137, 140
  premixed, 135, 136, 139
flammability, 138, 139, 177
flow, 1–3, 5, 8, 9, 13, 51–56, 75, 76, 79, 80, 82–84, 90, 91, 96, 97, 99, 100, 109–115, 117–119, 123, 125, 127, 128, 137, 184, 190, 192, 197–199, 205, 206, 211, 212, 214, 215, 218, 223
  rate, 2, 3, 83, 84, 91, 114, 197, 198, 205, 206, 214, 215, 218
flux
  heat, 9, 56, 198, 224, 231
friction, 5, 29, 81, 95, 98, 121, 122, 214
fuel, 25, 28–30, 34, 44, 45, 61, 67, 72, 74, 78, 79, 81, 83–85, 91, 93, 97–99, 111, 116, 119–121, 125, 130, 133, 135–145, 148, 150, 151, 166, 169, 171, 172, 176, 197, 203–205, 207, 209, 216–218, 221–224, 228, 229
  cell, 222, 223

G, H

gap, 128–131
gas, 14–17, 19, 20, 24, 29, 30, 32–35, 37, 39, 42–61, 63–68, 74, 83, 86–88, 91, 93, 95–98, 101, 103, 116, 120–124, 128, 135–138, 142, 148, 150, 151, 156, 160, 166, 167, 169, 170, 180–182, 188, 190, 192, 194, 195, 205–210, 212, 213, 216–222, 224, 228–231
combustible, 177
combustion, 195, 212
  ideal, 14, 16, 17, 19, 30, 32–34, 37, 39, 43, 45–48, 51, 55, 56, 63, 65, 87, 91, 95, 97, 98, 142, 160, 166, 167, 169, 170, 181, 182, 190, 192
  perfect, 34, 59, 64, 182
  semi-ideal, 16, 17, 44–47, 50, 51, 54, 59, 145, 150, 153, 156, 159, 160
  van der Waals, 20, 45
head loss, 25
heat, 1–9, 12, 15, 18–21, 30–33, 35, 37, 40, 42–49, 51, 53, 54, 56, 57, 58, 60, 66–68, 87–89, 92, 93, 95–97, 99–102, 110, 111, 119, 121, 122, 126, 127, 132, 134, 141, 145, 146, 149, 153, 154, 163, 167, 170–173, 179–184, 187, 188, 191, 193–195, 197–206, 209–212, 214, 216–219, 221–232
  capacity, 19, 44–46, 92, 153, 167, 181–183, 198
  exchanger, 89, 101, 102, 111, 179, 188, 193, 211, 214, 219, 221, 225
  production, 54
  transfer, 8, 56, 67, 68, 87, 126, 194, 198, 212, 224, 231
heater, 193, 194
heating value, 145–150, 174, 176, 197, 204

I, J

ignition, 23–25, 27, 28, 41, 57, 138–141, 177, 217
  compression, 24, 28, 41, 57, 139–141, 217
  spark, 25
incidence, 80, 82, 117
injection, 25, 27, 28, 58, 231
intake, 23, 25–28, 30–32, 34–36, 43, 47, 48, 51–54, 57, 59–65, 67, 74, 89, 91, 96, 105, 111, 112, 139, 140, 189–191, 193, 198, 219
intensive (quantity), 157
intrados, 80, 123
irreversibility, 6, 9, 11, 168, 201

irreversible, 7, 8, 12, 13, 21, 110, 190, 201
isobar, 66, 88, 91–93, 95, 97–101, 103, 126

L, M

law (thermodynamics)
    first, 2–4, 6–8, 10, 12, 21, 34, 40, 48, 49, 53, 55, 60, 65, 98, 145, 146, 152, 191, 197, 200, 201, 204
    second, 5, 8–10, 12, 21, 200, 201
    third, 173
lift, 54, 55, 78, 80–83
load, 25, 27, 30, 31, 33–36, 40, 43, 184, 203–205, 212, 215, 218, 220, 221, 230
mass, 1–3, 5, 14, 15, 25, 30, 32–35, 37, 39, 42, 47–49, 51–53, 55–61, 63–65, 75, 76, 78, 79, 81, 83, 84, 87, 90, 91, 99, 109, 111, 112, 114, 116, 118, 129, 131, 132, 141, 142, 144, 145, 148, 150, 180–182, 184, 189–192, 197, 198, 206, 212, 214, 215, 218, 219, 231
    molar, 87, 148
material body, 5
mechanism (chemical), 164, 165
mixture
    lean, 143, 144, 150, 152
    rich, 143, 144, 150, 153, 154, 222
    stoichiometric, 44, 45, 120, 154

N, O

nozzle, 55, 56, 74, 76, 78, 86, 88–90, 98, 105, 109, 111, 124, 125, 129
operating, 5, 6, 23, 26–30, 32, 43, 44, 66, 80, 86–88, 91, 93, 96, 111–113, 125, 140, 180, 181, 187, 188, 195, 200, 202–205, 212, 217, 221, 223, 226, 228
oxidizer, 72, 78, 134, 136, 142, 205
oxygen, 18, 20, 34, 72, 87, 133, 134, 141–145, 152, 155, 161, 165, 174, 175

P, R

piston, 3, 10, 23, 24, 26–32, 37, 47, 48, 52–54, 56, 57, 59, 61, 65, 71, 72, 91, 139, 140, 146, 186, 190, 200, 201, 210, 224, 225
polar curve, 82
pollutant, 120, 176
power, 4, 9, 13, 23–25, 28, 61, 67, 70–72, 84, 85, 106, 116, 130, 134, 165, 186, 195, 197–199, 203–207, 209, 211, 212, 214–218, 220, 227–232
principle of operation, 3
propeller, 71, 73–75, 89, 99
propulsion, 69, 72, 78, 79, 83, 89, 101, 116, 132, 135
reaction (chemical), 44, 133, 142, 144, 145, 147, 152, 157, 159–161, 165, 173
reaction rate, 158
reactive mixture, 156, 159
recovery, 67, 102, 179, 181, 182, 184, 193, 195, 197, 202–207, 209, 212, 215, 216, 218, 219, 221, 225–227, 230–232
reference state, 11, 12, 14, 21, 65, 66, 91, 147, 167, 168, 170
regenerating, 180–184, 186, 194
regenerator, 11, 12, 14, 21, 65, 66, 91, 147, 167, 170
reversible (cycle), 6, 8, 10, 200, 201
rocket, 71, 72, 78, 79, 135

S, T

section, 9, 44, 47, 50, 74, 83, 89, 93, 99, 104, 114–116, 118, 126, 137, 145, 146, 170, 186, 192, 203
stage, 103, 110, 116–118, 125, 129, 219, 229
stall point, 82
standard, 88–91, 93, 104, 116, 138, 147–149, 151, 152, 155, 160, 163, 173, 175, 176, 206, 216
station, 89, 90, 109, 113, 115, 121, 128
stoichiometry, 44, 45, 120, 142, 147, 150, 154, 157, 158, 161
supercharging, 61, 62, 66, 67, 197, 205, 214
surge, 80, 101, 114, 119
system
    closed, 2, 7, 10–12, 26, 30, 32, 40, 48, 49, 52, 63, 65, 172, 202
    open, 2–4, 12, 94, 200, 202
    propulsion, 83, 89, 101, 132
    thermodynamic, 2, 6, 32, 52, 60, 84, 191, 204

temperature
  explosion, 150
  flame, 152–155, 170, 171
thermoacoustics, 224, 225
thermoelectricity, 223, 224
thrust, 71, 75, 76, 78, 79, 82–85, 99, 106, 108, 111, 114, 124, 127
transformation
  adiabatic, 8, 16, 17, 45, 64, 95, 170
  isochoric, 181
  isothermal, 92, 150
turbofan, 77, 78, 90, 103

turbojet, 72–78, 83–86, 89, 93, 96, 97, 99, 104, 110, 111, 119, 122, 125, 128
turboprop, 73–76, 99

V, W

valve, 25–28, 30–32, 34, 36, 43, 50, 54, 55, 61–65, 131, 189–191
volume
  clearance, 24, 47, 188–191, 194
  displaced, 24, 57, 190
  ratio, 28, 47, 139, 182, 184, 190
wing, 80–82, 114, 132

Printed and bound by CPI Group (UK) Ltd, Croydon, CR0 4YY
05/01/2023

03177944-0001